T0406612

Springer Series in
MATERIALS SCIENCE   **113**

Springer Series in
MATERIALS SCIENCE

Editors: R. Hull   R.M. Osgood, Jr.   J. Parisi   H. Warlimont

The Springer Series in Materials Science covers the complete spectrum of materials physics, including fundamental principles, physical properties, materials theory and design. Recognizing the increasing importance of materials science in future device technologies, the book titles in this series reflect the state-of-the-art in understanding and controlling the structure and properties of all important classes of materials.

Sanat K. Chatterjee

Crystallography and the World of Symmetry

With 104 Figures

 Springer

Professor Sanat K. Chatterjee

National Institute of Technology, Physics Department,
Mahatma Gandhi Avenue, Durgapur-713209, West Bengal, India
E-mail: sanat_chatterjeein@yahoo.com

Series Editors:

Professor Robert Hull

University of Virginia
Dept. of Materials Science and Engineering
Thornton Hall
Charlottesville, VA 22903-2442, USA

Professor R. M. Osgood, Jr.

Microelectronics Science Laboratory
Department of Electrical Engineering
Columbia University
Seeley W. Mudd Building
New York, NY 10027, USA

Professor Jürgen Parisi

Universität Oldenburg, Fachbereich Physik
Abt. Energie- und Halbleiterforschung
Carl-von-Ossietzky-Strasse 9–11
26129 Oldenburg, Germany

Professor Hans Warlimont

Institut für Festkörper-
und Werkstofforschung,
Helmholtzstrasse 20
01069 Dresden, Germany

Springer Series in Materials Science ISSN 0933-033x

ISBN 978-3-540-69898-2   e-ISBN 978-3-540-69899-9

Library of Congress Control Number: 2008930151

Typesetting: Data prepared by SPi using a Springer TeX macro package
Cover concept: eStudio Calamar Steinen
Cover production: WMX Design GmbH, Heidelberg

SPIN: 12252381   57/856/SPi
Printed on acid-free paper

9 8 7 6 5 4 3 2 1
springer.com

To
Prajna, my wife
and
Baishampayan and Baijayanti,
my son and daughter
who always aspire to find a symmetry
between
thoughts and actions

Preface

From the time unknown, we are appreciating the symmetry present in nature. Starting from the beautiful wings of the butterfly and the colourful flowers the world of symmetry includes the glittering gem stones. People developed the methods of gems cutting and polishing and irrespective of the sex; gems had taken their place in ornaments.

Every man being vertebrate walks similarly on foot and so a symmetry or in variance exists in the mode of their walking, but there still exists something distinctive and characteristic in every man even in walking so that we can easily recognize in dark a known individual by observing simply his style of walking. There lies the asymmetry or a deviation from total invariance. The art of finding the wonderful symmetries that exists in nature is amazing and also the finding of asymmetry so intrinsically associated with it is probably more amazing.

The book has a beginning from the concept of crystal pattern, the lattice, different crystal lattices and the Space groups (Chaps. 1–8). A rather brief passage through this concept is made by discussing the different methods for the determination of the structure of crystals.

The book is basically intended to introduce the well known topic "Crystallography" from a different angle and so the topic has been introduced from the very initial concept of patterns and the symmetry present in crystals to the world of symmetry with the main aim to induce the general readers to keep their eyes open a bit wider to observe the ever attracting symmetry that the nature has left before us to apprehend and appreciate (Chap. 9). An emphasis is drawn on the symmetry present in the natural and the man-made objects, from flower, animal bodies to the ancient as well as the present day man-made engineering structures. From paintings to the nature and the laws of nature, from the concept of symmetry to asymmetry which is present along with the symmetry in nature and in some crystalline form of matter is also introduced.

Imperfect crystals are close to the asymmetric stage of matter but being variant in the characteristics of symmetrical state, they demonstrate some important properties that an ideally perfect symmetrical state fail to give. Therefore, attention is then shifted from single crystal state to polycrystalline state and some of their characteristic properties.

The concepts and characteristics of semi-crystalline states like liquid crystals, quasi crystals and finally the nano crystalline states are discussed in Chap. 10. While discussing the relevant aspects of this asymmetrical state of matter due attention has been given to discuss the properties and particularly the peculiarities of their structure dependant properties which are only possible to exist because of their deviation from perfect geometrically symmetrical arrangements of the constituents. The entire development is correlated with the symmetries and also the asymmetries present in matter and the laws that explain their characteristics.

Though the outline of this book is designed to serve those having only the basic introduction to mathematics, a brief introduction to the diffraction theory of perfect periodic to aperiodic structures is given in Appendix A. Some solved problems are also given in Appendix B to help students.

I express my thankfulness to Dr. M.K. Mandal for his help in proof reading. I express my gratefulness to Prof. L.S. Dent Glasser, Department of Chemistry, U. Aberdeen (Rtd) for allowing to reproduce some of the diagrams from her book "Crystallography and its applications." I also thank my students whose interest in the subject and interactions within and outside the class room gave me the idea to undertake this venture. If the content and presentation of this book satisfy their inquisitiveness, I would consider this venture as successful.

Durgapur, India   *Sanat Kumar Chatterjee*
July 2008

Contents

1

Pattern

1.1 Pattern: An Introduction

Let us begin with the question: What is a pattern? The answer to this question is as much objective as it may be subjective. From the days unknown, the human race have started studying and appreciating the regular periodic features like movement of stars, moon, sun, the beautiful arrangement of petals in flowers, the shining faces of gems, and also the beautiful wings of a butterfly. They have constructed many architectural marvels like tombs, churches, pyramids, and forts having symmetries, which still attract tourists. The regularities observed in nature either in the worlds of plants, animals, or inanimate objects are patterns and people get startled by observing them and thrilled by inspecting them. That may be the beginning of the study of pattern. Every music or song has two aspects: the tal or the tune and the verse of the song. The composer composes the tune on the verse of the song made by the lyric. This composition must satisfy certain harmonic conditions and the listener of the song has the right to appreciate a song or reject it. This judgment is more subjective than objective as to some listener some songs are very pleasing and appreciating where as to others its appeal may not be that much deep rooted. The appeal of a song sometimes also changes with time; new tunes come in the front and the old one are either rejected or forgotten. A pattern has much similarity with the sense of order, harmony that a song brings forward to us. It is also composed of two aspects: The first one like the verse of the song is the motif or the object which is the constituent of the pattern which can be compared with the song. The other is the order of the arrangement of the motif, which has similarity with the tune or the music. The most common example of a geometric pattern is a printed cloth or a wall paper. In a printed cloth the motif can be any object, it can be any geometrical figure, a leaf or a bud or anything. But the conventional sense of pattern can only be created if these motifs are either all similar, regular, or may be even a combination of more than one type. Now about the arrangement of these motifs in two-dimensional

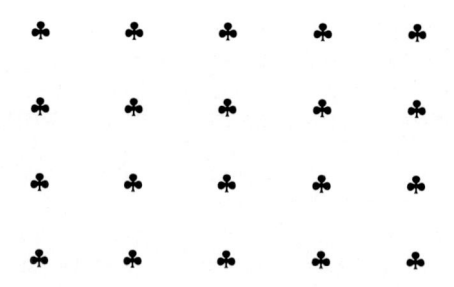

Fig. 1.1. Same motif, a perfect pattern

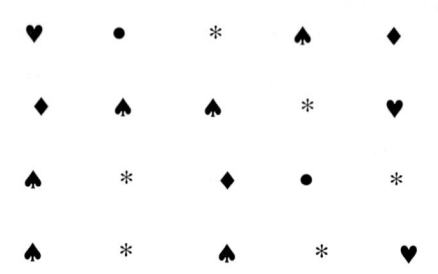

Fig. 1.2. Regular arrangement of random motifs, not a pattern

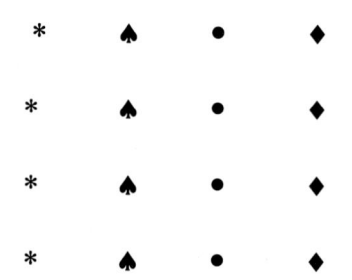

Fig. 1.3. The motifs are different but they bear a constant regularity in their arrangement, and so it constitute a pattern

printed cloths must also satisfy an order of their arrangement both in their translation or position and after orientation [1].

In Figs. 1.1 and 1.2, two arrangements of motifs are created such that in one all the motifs are similar and in the other, they are different. Now the question is, which one of these two appears more soothing to our eye? There lies perhaps the subjectivity of the problem. But if we do not only restrict ourselves to the more common geometric senses, then also the first one (Fig. 1.1) appears definitely more soothing to the eye and so can be accepted as a pattern and whereas the latter is not at least as in Fig. 1.1.

In Fig. 1.3 motifs are not all same row-wise but they are same column-wise.

Note: The scheme of repetition or the mode of arrangement of motifs is the same in all figures. The only difference is that in Fig. 1.1 the motifs are similar and in Figs. 1.2 and 1.3 they are different. The pattern in Fig. 1.1 is more regular than that in Fig. 1.2 and also in Fig. 1.3, but Fig. 1.3 is more "pattern-like" than Fig. 1.2.

Conclusion: To constitute a pattern, the motifs are either to be same or should be regularly arranged in the same scheme of repetition even if they are different. The scheme of repetition comprises of position and orientation of the motifs.

Now, if the motifs are identical but the mode of their arrangement, that is, the scheme of their repetition is changed then the pattern will also be changed. This is shown in Figs. 1.4 and 1.5.

Note: In both Figs. 1.4 and 1.5, the patterns are created but they look different because their schemes of repetition are different. Again in Fig. 1.6 though there is regularity in arrangement, it looks less pattern-like than Fig. 1.7 as the color of the motifs is random in Fig. 1.6, whereas the motifs are systematically colored in Fig. 1.7 and so it appears more pattern-like. In Fig. 1.8 each motif are similar but randomly oriented but positioned in regular order, and in Fig. 1.9 both are maintained and so it is a regular geometric pattern. Figure 1.10 shows another pattern though motifs are oriented but in symmetrical right-hand screw order. This can be understood in more elaborate

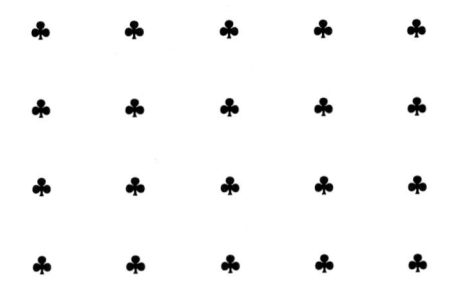

Fig. 1.4. A pattern with one scheme of repetition

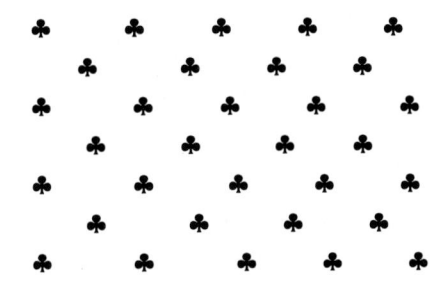

Fig. 1.5. A pattern with different scheme of repetition

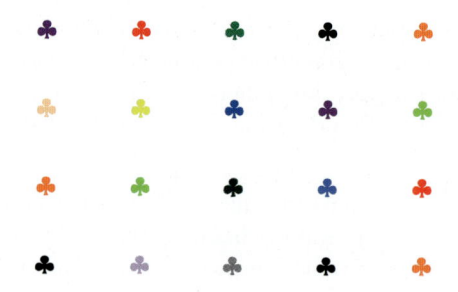

Fig. 1.6. Regular arrangement of randomly coloured motifs

Fig. 1.7. Regular arrangement of symmetrically coloured motifs

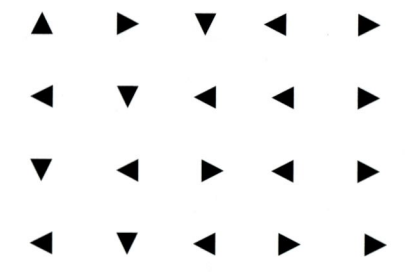

Fig. 1.8. Same motifs, symmetrically placed but randomly oriented

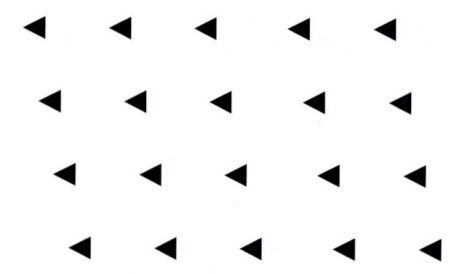

Fig. 1.9. Same motifs symmetrically placed and oriented

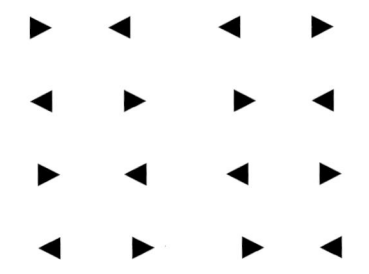

Fig. 1.10. Motifs symmetrically placed and also oriented

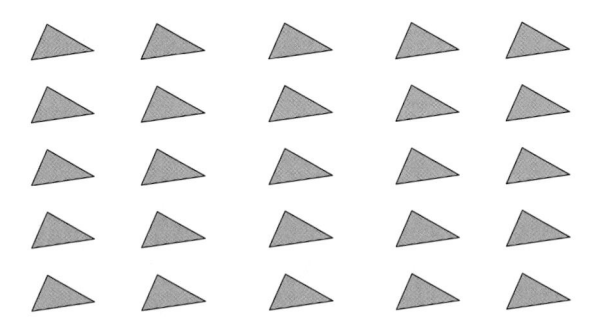

Fig. 1.11. Motifs are triangles and each is placed in perfect symmetrical position to constitute a pattern

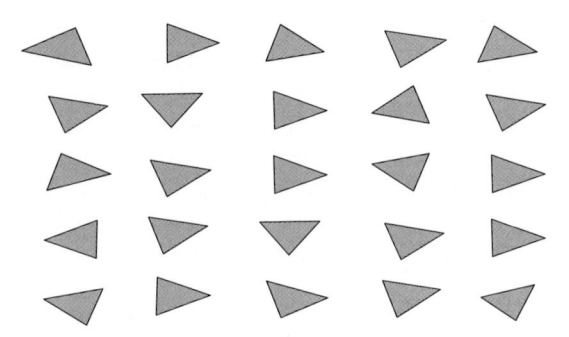

Fig. 1.12. Motifs are triangles and are same as Fig. 1.11 and they are placed in perfect symmetrical positions but are randomly oriented about their positions and so it loses the characteristics of the pattern

way from Figs. 1.11 and 1.12, where in the motifs are triangles and they are differently placed so far as their orientation is concerned but in symmetrical positions.

Conclusion: The change in scheme of repetition either in position or orientation changes the patterns and even loses the sense of pattern if there is no regularity in their orientation, though the motifs remain same and they are placed at equal intervals.

Therefore, a pattern must possess two characteristics, the regular motifs stationed at sites that obey a certain scheme of repetition [1, 2]. Any change of motif or the change of the repetition scheme both in their positions and orientation changes the pattern. So to study a pattern these two aspects are to be looked into and studied. It is more convenient to start with the study of different possible schemes of repetition that exist in their position rather than with the orientation symmetry, if there is any, in the motifs. The former class of study may be categorized as Macroscopic whereas the latter may be called as Microscopic. To begin with the investigation on scheme of repetition, it should be appreciated that the regularity of the scheme of repetition including both macro and micro is the essence of a perfect pattern [3]. This regularity generates a sense of symmetry, which is later described as Macroscopic and Microscopic. It is better to start with the symmetry present in the scheme of repetition, that is, with the symmetry operations.

Note: Two bodies or configuration of bodies (i.e., Pattern) may be called symmetrical if and only if they are indistinguishable in all respect. The symmetry operations are those operations which when performed on a pattern, the pattern returns back to its state of self-coincidence or invariance.

Thus the total identification of the symmetry operations that can be performed to bring a pattern in to its self coincidence, gives the knowledge of the symmetry and the scheme of repetition present in the pattern.

This should be appreciated that the entire knowledge of symmetry or scheme of repetition present in the pattern though include the sense of motifs present, it would be much beneficial at least to start with if we consider the symmetry present within the sites of the motifs only (Macroscopic), and when these are known, then the symmetry operations including the motifs (Microscopic) can be considered. Therefore, let us start with the symmetry present within the sites where the motifs are to be placed to generate the patterns.

These sites of the motifs, which are represented by simple geometrical points, have a special significance and they are known as "Lattice."

Conclusion: The Lattice is the sites of motifs where they can be placed to generate the pattern. It can be two-dimensional regular arrays of points for two-dimensional patterns or it can be three-dimensional arrays of points for three-dimensional patterns. Therefore, the lattice bears the knowledge of the scheme of repetition and when the motifs are placed in the lattice sites the entire pattern takes the shape and changes whenever the orientation of the motifs takes their role to play. If the order of this orientation of the motifs is maintained in some way or other, it retains the pattern characteristics of being geometrically symmetric otherwise not.

1.2 Summary

1. A pattern has two constituents: one is the *motif* and the other is *the scheme of repetition* of the motif.
2. The scheme of repetition of motifs again has two aspects: Positions in the lattice sites (macroscopic) and the orientation (microscopic) in their respective lattice sites.
3. A pattern has two aspects: one is the structure of the motif and its schemes of repetition and the other is its pleasant visual effect. This is the after effect when these two aspects are followed. Therefore, both these two aspects, that is, the symmetry and the visual effect are important and are usually supplementary to each other.
4. When the first aspect is not at all or is only partially followed, it is not necessary that the assembly of the motifs will not satisfy the second aspect of it, that is, the visual pleasure and will fail to constitute a pattern. It is then an incidence of an exception in the geometric rule of pattern, that is, asymmetry in symmetry, which is abundantly present in nature.

References

1. N.F. Kenon, *Patterns in Crystals* (Wiley, New York, 1978)
2. L.S. Dent Glasser, *Crystallography and its Applications* (Van Nostrand Reinhold Co. Ltd, New York, 1977)
3. L.V. Azaroff, *Introduction to Solids* (McGraw-Hill, New York, 1960)

2

Lattices

2.1 Plane Lattice

The two-dimensional infinite array of geometrical points symmetrically arranged in a plane where the different motifs may be placed to create the patterns is known as plane lattice. Figures 2.1 and 2.2 are the examples of plane lattices, where the neighborhood of every point is identical. Figures 2.1 and 2.2 are two types of plane lattices, where motifs are to be placed to generate the desired patterns [1].

Note: The plane lattice of a two-dimensional pattern is that array of geometric points which specifies the scheme of repetition that is present in the pattern. There is no difference between the neighbors of one lattice site from any of its neighborhood. Actually the situation becomes so much identical that if the attention from one lattice point is removed, it then becomes impossible to identify and locate the same lattice point.

Note: The unit translation is the distance shown by arrow between adjacent points along any line in a lattice. This distance taken between any adjacent point is a, and d is the distance of separation between any line drawn through the points in the lattice (Figs. 2.3 and 2.4).

The product of a and d, that is, $a \times d$ is the total area associated with each lattice point and the inverse, that is, $1/(a \times d)$ is the number of points in each unit area of the lattice. The unit cell of the plane lattice is a parallelogram of two unit translations with lattice points at the corners and is the representative of the lattice, that is, the lattice is constituted by repeating this unit cell in two directions of the plane lattice.

This unit cell is completely specified by the lengths of the edges known as unit translations and the angle between them. So, the unit cell is the building block of the lattice (Fig. 2.5) [1–3].

Now as this unit cell is the building block of the pattern, it may be thought that there can be infinite number of patterns even for same motifs for different

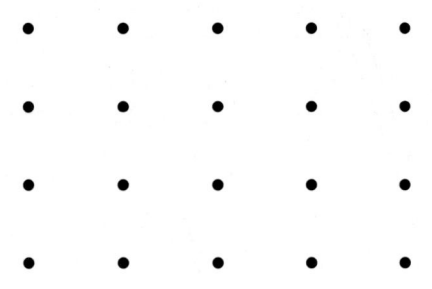

Fig. 2.1. Plane Lattice (Type I)

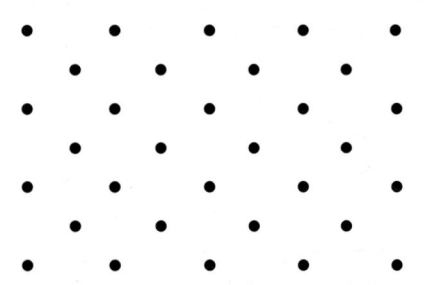

Fig. 2.2. Plane Lattice (Type II)

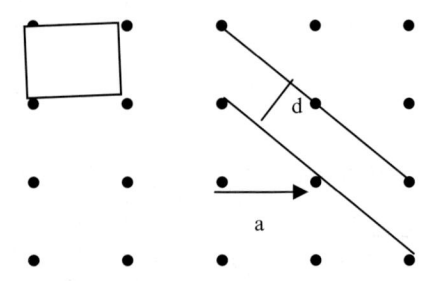

Fig. 2.3. Distances between lattice sites and lattice rows (Type I)

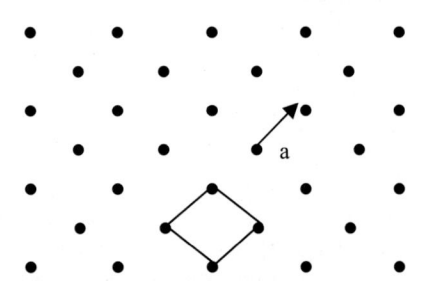

Fig. 2.4. Distances between lattice sites and rows (Type II)

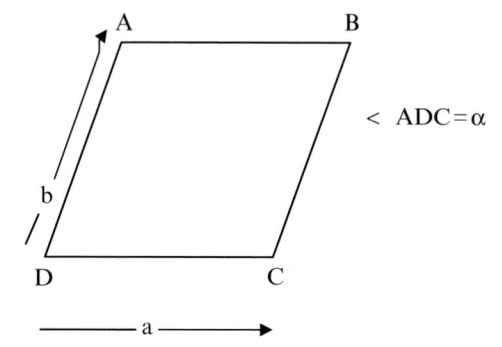

Fig. 2.5. A unit cell, ABCD the building block of the lattice. Two sides are of lengths a and b and α is the angle between them

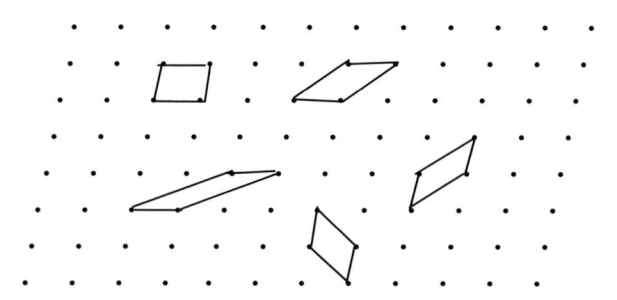

Fig. 2.6. Plane lattice structure. There are possibilities of different shapes of unit cells, each having different sets of a and b and also the angles between them, but each one preserve one condition: they are formed by joining only the corner sites

unlimited values of a and b, but the actual fact is that the pattern types are dependent not on the value of a and b but whether $a = b$ or $a \neq b$ and whether $\alpha = 90°$ or $\alpha \neq 90°$ but equal to $120°$. These different conditions can only lead to different types of plane lattices. Based on this observation it can be concluded that there are five different types of plane lattices out of six possible unit cells as a pentagon cannot act as a building block for a continuous two-dimensional pattern [2,3].

(1) Square [$a = b$ and $\alpha = 90°$]
(2) Rectangle [$a \neq b$ but $\alpha = 90°$]
(3) Rhombus [$a = b$ and $\alpha \neq 90°$]
(4) Pentagon [$a = b$ and $\alpha = 108°$]*
(5) Hexagon [$a = b$ and $\alpha = 120°$]
(6) Parallelogram [$a \neq b$ and $\alpha \neq 90°$].

   *This unit cell is not considered as a building block.
   Figures 2.6 and 2.7 show the unit cells of five possible plane lattices.

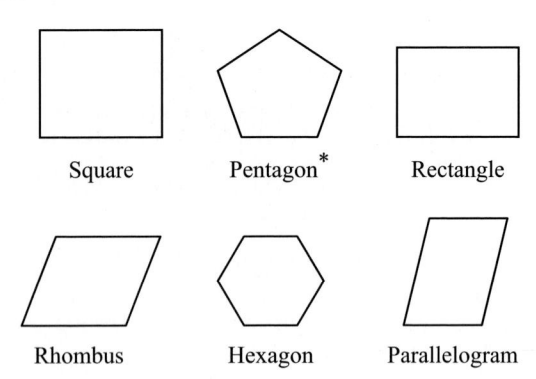

Fig. 2.7. Different possible shapes of unit cells of plane lattices

The question may arise why a pentagonal unit cell is not generally possible? The answer to this interesting question may be obtained from the consideration of two aspects of the pattern. The one is the unit cell consideration and the other from the symmetry consideration, which will be discussed later. From the first consideration, it can be said that as the unit cells are building block of the pattern, by repeating the pentagon in a plane it is not possible to construct a plane lattice. This aspect as mentioned will be taken up in a later chapter.

2.2 Space Lattice

If we add one more dimension to the plane lattice and arrange the geometrical points also in the third axis obeying the order, then it constitutes a space lattice. The unit cell of the space lattice is the three-dimensional building block of the space lattice, which when are arranged in three directions obeying the order of repletion will constitute the space lattice. The three axial lengths and the angles between them then are required to specify the unit cell. Figure 2.8 shows the unit cell of a space lattice.

2.3 Lattice Planes and Miller Indices

Unlike plane lattice where the lines joining lattice points are lines, here in space lattice it generates a plane known as lattice plane. As it is important to know these lattice planes and as they preserve their individuality in much respect, each such lattice planes are identified by indices known as Miller indices, *hkl*. These indices are given to a plane by the following procedures [3, 4]:

(a) Measure the intercept that the plane makes on the axes of the lattice
(b) Divide the intercept by the appropriate unit translation

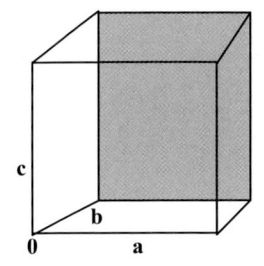

Fig. 2.8. *a, b,* and *c* are unit translations and α, β, and γ are the angles between *b* and *c, a* and *c,* and *a* and *b,* respectively

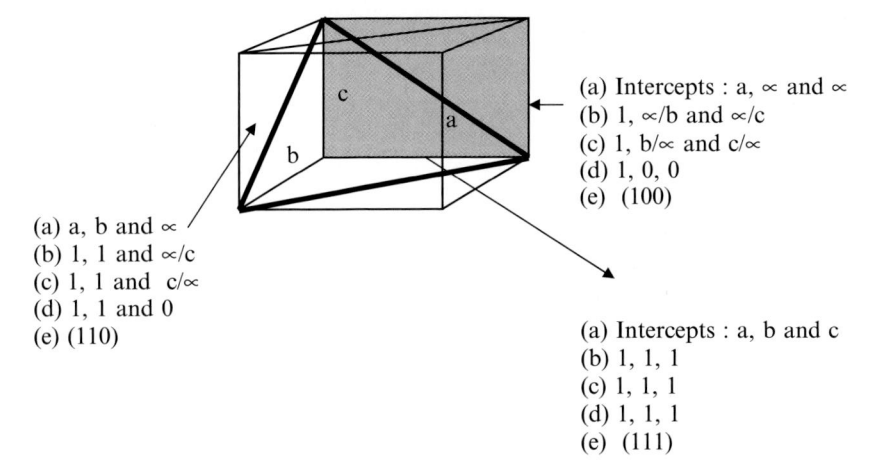

(a) Intercepts : a, $\propto$ and $\propto$
(b) 1, $\propto$/b and $\propto$/c
(c) 1, b/$\propto$ and c/$\propto$
(d) 1, 0, 0
(e) (100)

(a) a, b and $\propto$
(b) 1, 1 and $\propto$/c
(c) 1, 1 and c/$\propto$
(d) 1, 1 and 0
(e) (110)

(a) Intercepts : a, b and c
(b) 1, 1, 1
(c) 1, 1, 1
(d) 1, 1, 1
(e) (111)

Fig. 2.9. The Miller indices of some planes

(c) Invert the dividends
(d) Rationalize the inverted dividends
(e) Place the rationalized numbers in round brackets

Figures 2.9 and 2.10 explain the procedures as mentioned above.

The sets of planes (*hkl*) are noted by {*hkl*}. The {110} family of planes comprises six planes, that is, (110), (101), (011), ($\bar{1}\bar{1}$0), ($\bar{1}$0$\bar{1}$), and (0$\bar{1}\bar{1}$). The bars as usually signify the intercept in the negative side of the axes. All these planes belong to the same class and together they are represented by {110}.

2.4 Lattice Directions

Like the indices of a lattice plane, the lattice directions are also important aspects to be known. They are also noted by Miller indices and are done as follows:

(a) Measure the coordinates of any point on the direction to be named or indexed

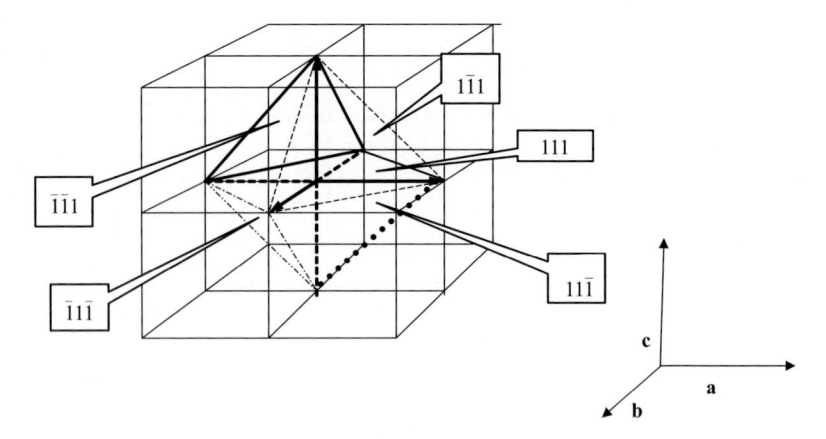

Fig. 2.10. Miller Indices of some more planes with axes showing separately

(b) Divide the coordinates of the point by appropriate unit translations
(c) Rationalize the dividends
(d) Place the rationalized dividends in a square bracket

Note: [uvw] refer to any direction in a space lattice. [$\bar{1}11$] and [$1\bar{1}\bar{1}$] are opposite senses of the same line [3, 4].

In space lattices in addition to the unit translational distances called vectors like *a*, *b*, and *c* and angles as α, β, and γ between them the distances between different lattice planes designated by their respective Miller indices are also important. These distances are known as interplaner distances and are designated by '*d*' spacings. The derivation of their expressions is done in a later chapter after the introduction of crystal classes.

2.5 Summary

1. A lattice is an array of geometrical points either in two or in three dimensions to make plane or space lattices.
2. This array of points shows the order in the scheme of repetition.
3. This scheme of repetition is maintained although the lattice, that is, every region of a lattice is exactly identical with other.
4. The different planes and also the directions on which these lattice points lie are designated by indices known as Miller indices.

References

1. N.F. Kenon, *Pattern in Crystals* (Wiley, New York, 1978)
2. L.V. Azaroff, *Elements of Crystallography* (Mc Graw Hill, New York, 1968)
3. C.S. Barrett, T.B. Massalski, *Structure of Metals* (Mc Graw Hill, New York, 1956)
4. M.J. Buerger, *Elementary Crystallography, an Introduction to the Fundamental Geometrical Features of Crystals* (Wiley, New York, 1963)

3

Symmetry in Lattices

In discussing the unit cells of five different kinds in plane lattices, it has been understood that more a unit cell is complex, greater is the number of parameters required to specify that cell. This complexity of the unit cells is determined by the "symmetry properties" of the array of points in the different types of plane and space lattices.

The symmetry properties of a lattice are specified by one or more than one operation, which when are executed over a lattice point, the entire lattice can be constructed. Each point of the lattice is brought back to another identical point. It should be taken in mind that when we discuss only with lattices, the symmetry exists only due to the scheme of repetition, but when the motifs are placed the symmetry changes, but the symmetry in the lattice is the minimum symmetry present.

3.1 Symmetry Operations in Plane Lattices

3.1.1 Rotational Symmetry

In Fig. 3.1, ABCD is a parallelogram. If the diagram is rotated along X-Y axis clockwise or anticlockwise through angles $180°$ or $360°$, then the diagram returns to its self coincidence. The minimum angle through which it can be rotated to bring it into self coincidence is $180°$. If instead of a parallelogram or rectangle the diagram was a square, then the minimum angle through which it had to be rotated for self coincidence would be $90°$.

These rotation axes of symmetry are noted by $360/\theta = n$, where θ is the angle of rotation for self coincidence and n is known as "fold of rotation." Thus for parallelogram or for rectangle n should be 2 about X-Y and for square it should be 4.

(1) Rotational symmetry:

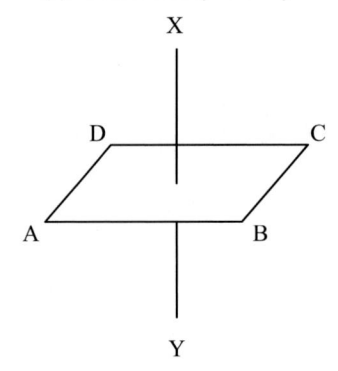

Fig. 3.1. Rotation axis of symmetry

Note: A n fold rotation axis of symmetry is a line about which the body or the pattern is transformed to self coincidence n times during a 360° rotation.

The rotation axes of symmetry from onefold to sixfold rotation are shown in Fig. 3.2a–f and the details of rotation angles to bring the lattice into self-coincidence are also given in Table 3.1. It is evident from the above figures that onefold of rotation is present in any irregular shaped body; the fivefold rotation axis of symmetry is not possible in regular pattern as its unit cell, which is a regular pentagon, cannot be arranged even in a plane so as to make any plane pattern. There is some exception to this, that is, possibility of fivefold symmetry (Fig. 3.3). This aspect will be taken up in a later Chapter. Sevenfold or more than that will either fail to make a regular unit cell or they will be repetition of any of the one- to sixfold of rotations [1–3].

Note: The highest rotational symmetry that occurs in the plane lattice is the rotational symmetry of the plane lattice and fivefold of rotation is not possible.

However, there are some exceptions to this in quasi crystalline state of matter, which will be discussed in a later chapter.

3.1.2 Mirror Plane of Symmetry

A mirror plane (Hermann–Mauguin symbol: m) is any plane that divides the lattice into two halves so that one half is the mirror image of the other half across the plane. As a lattice is made up of regular points (sites of motifs), the number of mirror planes in a plane pattern may be less than the number in a plane lattice. This depends on the shape and symmetry of the motifs which make that pattern.

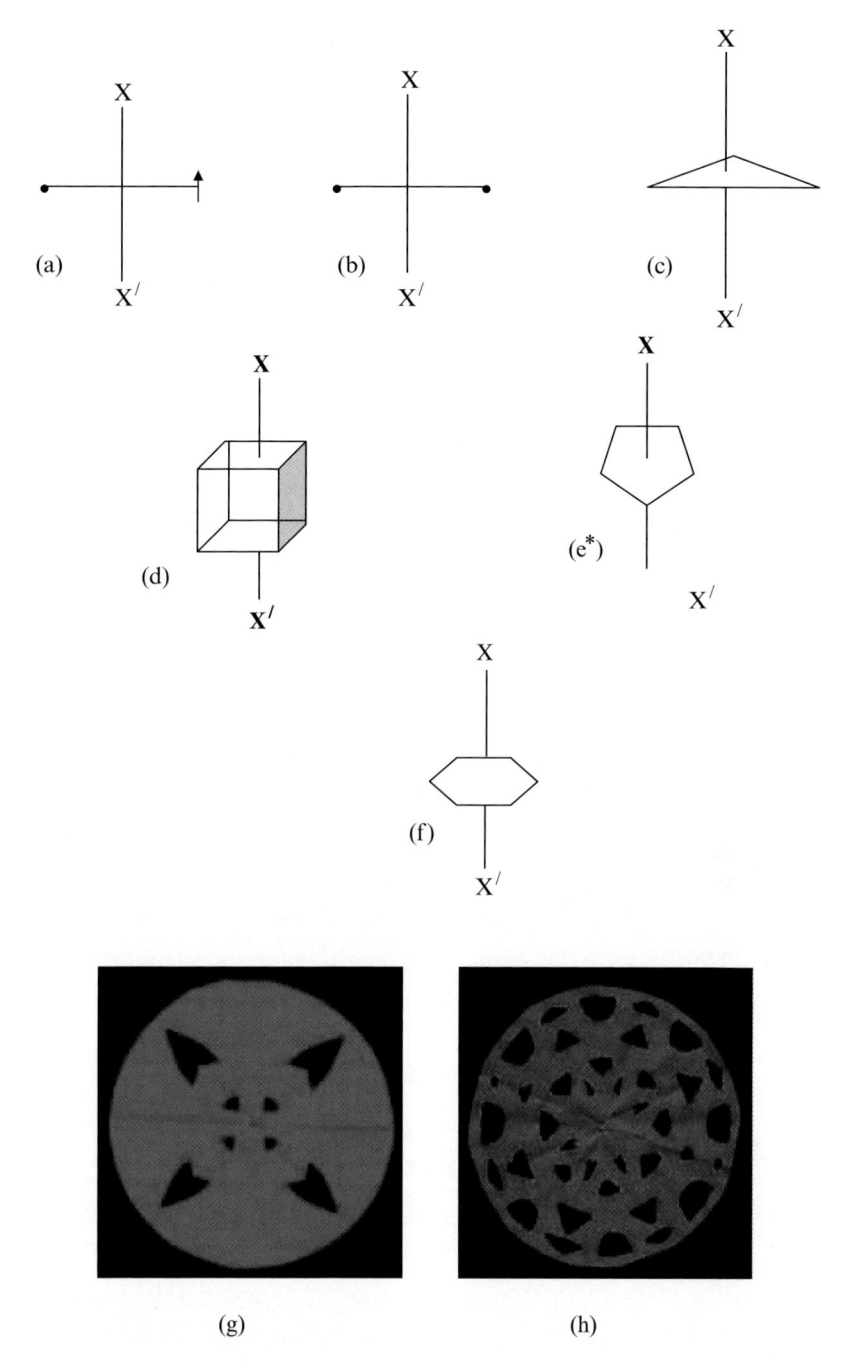

Fig. 3.2. (a–f) Rotation axis for $n = 1$, 2, 3, 4, 5, and 6 and **(g)** and **(h)** are, respectively, the patterns having fourfold and eightfold axis of symmetry

Table 3.1. Different rotational symmetries present in plane or space lattices

Sl. No.	θ, the rotation angle	$360°/\theta$	Symmetry axis	Numerical symbol (Hermann–Mauguin)
1	360°	1	Onefold	1
2	180°	2	Twofold	2
3	120°	3	Threefold	3
4	90°	4	Fourfold	4
5	72°	5	Fivefold	5*
6	60°	6	Sixfold	6

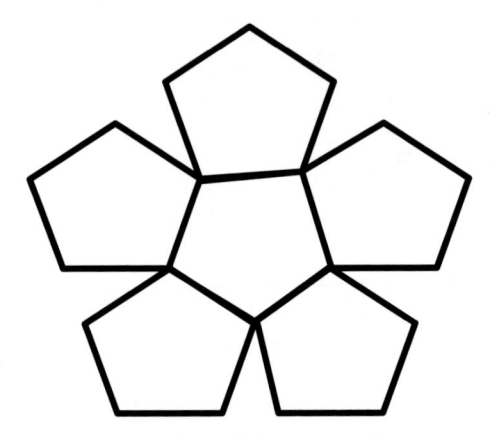

Fig. 3.3. It shows the impossibility of fitting regular pentagons having fivefold of rotation symmetry in condensed matter structure. Any such attempt will leave either voids or an overlap of the pentagons [2]

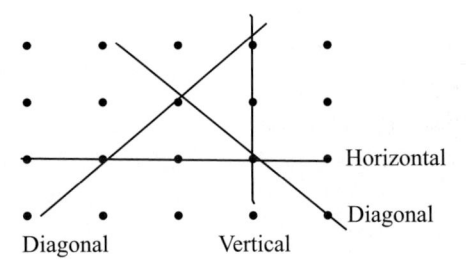

Fig. 3.4. Four mirror planes in a plane lattice

Note: In Figs. 3.4 and 3.5 the lattices are same but in Fig. 3.5 there is only one diagonal mirror plane because of the shape of the motif. Therefore, Mirror Plane also depends on the shape of the motifs. Therefore, the actual number of mirror plane of symmetry present in patterns depends on the shape of the motifs.

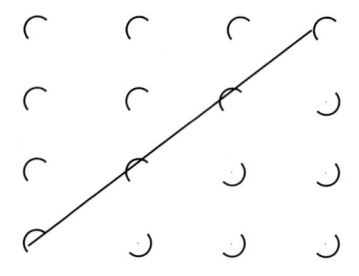

Fig. 3.5. One diagonal mirror plane in the pattern

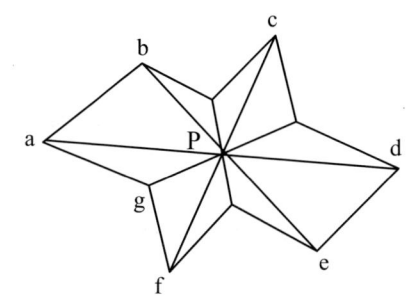

Fig. 3.6. An irregular polygon has a centre of symmetry at the point P. Each point is equidistant from P, from the corresponding point on the opposite side of centre point P. If every point within a body can be inverted through a centre then the body is transformed to self coincidence by the inversion

3.1.3 Centre of Symmetry

A body has a centre of symmetry (Symbol $\bar{1}$) if for every point in it there is an identical point equidistant from the centre but on opposite side in the inverted state (Fig. 3.6).

Note: The unit cells of all plane lattices have a centre of symmetry, but whether the unit cell of a pattern has a centre of symmetry depends on the shape of the motif.

3.2 Symmetry Operation in Space Lattices

3.2.1 Rotation Inversion (Rotary Inversion) Symmetry

A three-dimensional space lattice retains all the symmetry operations possible in plane lattices, that is, rotation axis, mirror plane, and centre of symmetries. In addition to these symmetries for reasons having one more dimension, a space lattice may have an additional symmetry operation, that is, rotation inversion symmetry or simply roto-inversion. This symmetry operation is the combined effect of pure rotation and inversion of the lattice point to bring it into self-coincidence.

They are the following:

1	(Onefold rotation)	+	$\bar{1} = \bar{1}^*$
2	(Twofold rotation)	+	$\bar{1} = \bar{2}$
3	(Threefold rotation)	+	$\bar{1} = \bar{3}$
4	(Fourfold rotation)	+	$\bar{1} = \bar{4}$
6	(Sixfold rotation)	+	$\bar{1} = \bar{6}$

Figure 3.7 demonstrates this rotation inversion symmetry operation in space lattices.

Note: The onefold roto-inversion ($\bar{1}$) and twofold roto-inversion ($\bar{2}$) may exist on a plane but higher roto-inversion symmetries must move out of the plane and so this symmetry is called a symmetry operation in space lattices.

Conclusion: The symmetry elements that may occur without any repetition in space lattices are as follows:

Rotation Axes: 1, 2, 3, 4, and 6
Centre of Symmetry: 1
Rotation Inversion Axes: 3, 4, and 6
Mirror plane: m (2)

Note: As $\bar{1}$, that is, onefold rotation and inversion is same as centre of symmetry and $\bar{2}$, that is, twofold rotations and inversion is also same as the mirror plane, they are not included as separate symmetry elements in space lattices.

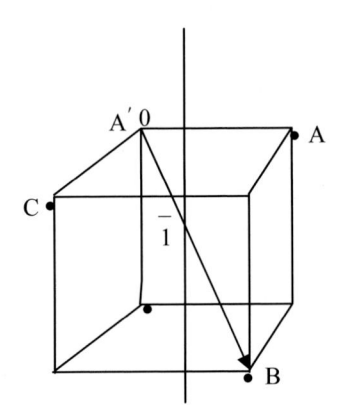

Fig. 3.7. Fourfold rotation inversion. A is given 90° rotation about the rotation axis *X-Y*. *Filled circle* changes to *open circle* (A to A′) then inversion about the centre to B

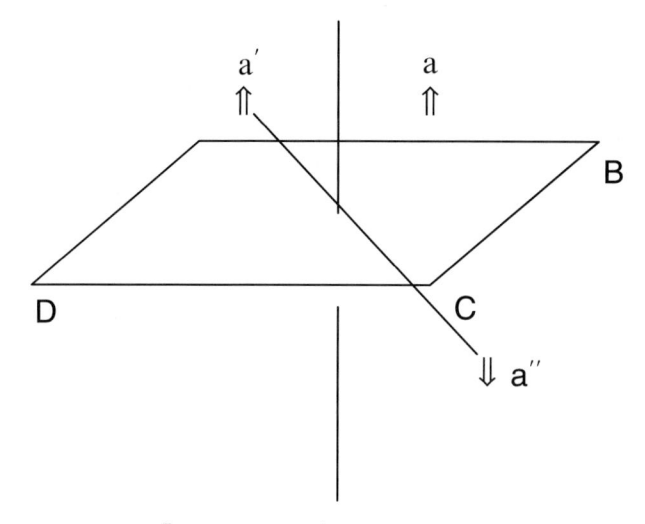

Fig. 3.8. Equivalence of $\bar{2}$ and m a to a′ by a twofold rotation and then inversion to a″. These entire operations are equivalent to a simple mirror reflection over the plane ABCD (a changes directly to a″)

Example: Figure 3.8 shows that a twofold rotation inversion is same as mirror plane. The motif ⇑ at position a has undergone a twofold rotation to position a′ and then subsequently by inversion to position a″, which can also be obtained by a simple mirror reflection over the plane ABCD [3].

3.3 Summary

1. Lattice, which is the array of points, the sites of the motifs, gives the knowledge of the particular scheme of repetition present.
2. The scheme of repetition is controlled by "symmetry operations," which when performed the lattice goes to an identical situation of existence, that is, the position of self coincidence or the position of invariance.
3. The scheme of repetition may or may not alter after the motifs are placed at the lattice sites as then the symmetry will also depend on the structure and nature of the motifs.

References

1. F.C. Philips, *An Introduction to Crystallography* (Oliver and Boyd, Edinburgh, 1971)
2. L.S. Dent Glasser, *Crystallography and Its Applications* (Van Nostrand Reinhold Co., New York, 1977)
3. W.L. Bragg, *The Crystalline State, Vol. I: A general Survey* (George Bell, London, 1933)

4

Crystal Symmetry (Crystal Pattern): I

4.1 Macroscopic Symmetry Elements

We have already introduced the space lattice and the symmetry elements that are possible in the space lattices. It has also been mentioned that when a pattern is constituted by placing the motifs in the lattice sites, these symmetry elements originally present in the space lattice of the pattern may be reduced due to the shape of the motifs; however, this is not true only when the motifs are isotropic in their shape. If the motifs are atoms or molecules then the pattern itself is a crystal. Therefore, the crystals can be regarded as three-dimensional patterns. When some of the symmetry elements present in the corresponding space lattices are also present in the crystal, they are named as macroscopic symmetry elements as they are manifested by the regular external faces of the crystals and as most of them can be studied by unaided eye by observing the single crystal faces. These macroscopic symmetry elements present in a crystal can be once again recalled as in Table 4.1.

The total macroscopic symmetry of a crystal or a pattern is then that collection of symmetry elements which is associated with the point at the centre of the crystal. Now, since every point of a space lattice is indistinguishable from every other point, the collection of elements defining the symmetry of the lattice must be associated with each and every point of the lattice. Thus, to completely specify the macroscopic symmetry of the lattice it is necessary only to specify the symmetry at any point in that lattice.

The collection of symmetry elements at any point of the lattice is termed as the *point symmetry* or the *point group of symmetry*.

Conclusion: The point group of symmetry of a crystal is that collection of macroscopic symmetry elements which occurs at every lattice point of the space lattice of the crystal taking into consideration that point group of a lattice may be different from the point symmetry of the actual crystal itself as a consequence of the shape of the motif (atoms or molecules).

Table 4.1. Symmetry Operations and Symbols

Sl. no.	Symmetry operation	Symmetry elements Hermann–Mauguin symbol	Total
1	Mirror plane	m or $\bar{2}$	1
2	Centre of symmetry	$\bar{1}$	1
3	Rotation axis	1, 2, 3, 4, 6	5
4	Rotary inversion	$\bar{3}, \bar{4}, \bar{6}$	3

Total number of macroscopic symmetry elements = 10

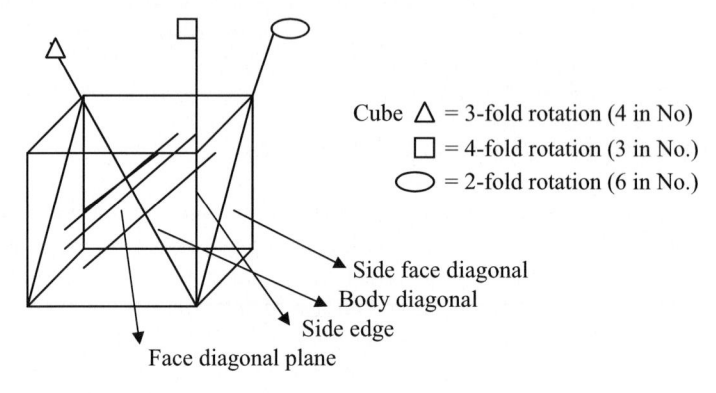

Cube △ = 3-fold rotation (4 in No)
□ = 4-fold rotation (3 in No.)
◯ = 2-fold rotation (6 in No.)

Side face diagonal
Body diagonal
Side edge
Face diagonal plane

Fig. 4.1. Different planes and rotation axes in cube. △ = threefold rotation (4 in number); □ = fourfold rotation (3 in numbers); ◯ = twofold rotation (6 in numbers)

A detailed analysis of the compatibility of the ten symmetry elements in all possible combinations is beyond the scope of this book. It can be analyzed that out of a large number of possible groupings of these ten symmetry elements, only 32 are compatible combinations of one or more of them and thus there exist only 32 *point groups* of symmetry (Fig. 4.1).

4.2 Thirty-Two Point Groups of Symmetries in Hermann–Mauguin Notations

1	1	A onefold rotation axis (no symmetry at all)
2	$\bar{1}$	A centre of symmetry only
3	2	A twofold rotation axis only
4	$m(\bar{2})$	A single mirror plane
5	$2/s$	A twofold rotation axis with a mirror plane at right angle to it

6	222	Three twofold rotation axes at right angles to one another
7	$2mm$	Two mirror planes at right angles to one another with a twofold rotation axis along the line of intersection of the mirror planes
8	$2/m\,2/m\,2/m$	Three mirror planes at right angles to one another with a twofold rotation axis along each of the three lines of intersection
9	3	A threefold rotation axis only
10	$\bar{3}$	A threefold rotary inversion axis
11	$3m$	Three mirror planes at 120° to one another intersecting along a threefold rotation axis
12	32	A threefold rotation axis passing at right angles through the intersection of three twofold axes at 120° to each other
13	$\bar{3}\,2/m$	Three mirror planes at 120° to each other intersecting along a threefold rotary inversion with three twofold axes at right angles to the rotary inversion and midway between the mirror planes
14	4	A fourfold rotation axis only
15	$\bar{4}$	A fourfold rotary invertor only
16	$4/m$	A fourfold rotation axis with a mirror plane at right angle to it
17	4 2 2	A fourfold rotation axis passing at right angles through the intersection of four twofold rotation axes at 45° to one another
18	$4mm$	Four mirror planes at 45° to one another with a fourfold rotation axis along the line of intersection
19	$4/m\,2/m\,2/m$	Four mirror planes at 45° to one another intersecting along a fourfold rotation axis; an additional mirror plane at right angles to the fourfold axis intersects the other mirror planes along four twofold rotation axes
20	$\bar{4}\,2m$	Two mirror planes intersecting along a fourfold rotary invertor with two twofold rotation axes at right angles to the rotary inversion and midway between the mirror planes
21	6	A sixfold rotation axis
22	$\bar{6}$	A sixfold rotary inversion
23	$6/m$	A sixfold rotation axis with a mirror plane at right angle to it
24	6 mm	Six mirror planes at 30° to one another intersecting along a sixfold rotation axis
25	6 2 2	A sixfold rotation axis passing at right angles through the intersection of six twofold rotation axes at 30° to one another

26	$\bar{6}\,2m$	Three mirror planes at 60° to each other intersecting along a sixfold rotary inversion, each mirror plane containing a twofold rotation axis at right angles to the rotary inversion axis
27	$6/m\,2/m\,2/m$	Six mirror planes at 30° to each other intersecting along a sixfold rotation axis, each mirror plane containing a twofold rotation axis at right angles to the sixfold axis and lying in a mirror plane also at right angles to the sixfold axis
28	23	Three twofold axes at right angles to one another and parallel to the edges of a cube with four threefold axes parallel to the body diagonals, i.e., four threefold axes at 70°32/to one another
29	$2m\,\bar{3}$	Three mirror planes parallel to the face of the cube, intersecting along three twofold axes parallel to the edges of the cube with four threefold rotary invertors parallel to the body diagonal of the cube
30	$\bar{4}\,3m$	Three fourfold rotary inversion parallel to the edges of a cube, with four threefold rotation axes parallel to the body diagonal and six mirror planes, each containing a face diagonal
31	4 3 2	Three fourfold rotation axes parallel to the edges of a cube with four threefold rotation axes parallel to the body diagonals and six twofold rotation axes parallel to the face diagonals
32	$4/m\,\bar{3}\,2/m$	Three fourfold rotation axes parallel to the edges of a cube with four threefold rotary invertors parallel to the body diagonals, six twofold rotation axes parallel to the face diagonals, and nine mirror planes, three of which are parallel to the faces and the other six containing a face diagonal

Note: The group of Sl. No. 32 is the most highly symmetrical of all the point groups of symmetry.

Now these 32 point groups, which are the only possible combinations of ten symmetry elements like mirror plane (m or $\bar{2}$), centre of symmetry ($\bar{1}$), rotation axis of symmetries (onefold, twofold, threefold, fourfold, and sixfold), and rotation (Rotary) inversion (threefold rotary inversion $\bar{3}$, fourfold rotary inversion $\bar{4}$, and sixfold rotary inversion $\bar{6}$), can be regrouped into seven different sets each having one common symmetry element. These seven different sets are listed in Table 4.2.

Table 4.2. Seven Crystal Systems

Set	Sl. Nos. grouped together with their symmetry elements	Common essential symmetry	Name of the set
A	1 and 2; 1, $\bar{1}$	One onefold rotation axis	Triclinic
B	3–5; 2, m, $2/m$	One twofold rotation axis	Monoclinic
C	6–8; 222, $2mm$, $2/m2/m2/m$	Three twofold rotation axes, perpendicular to one another	Orthorhombic
D	9–13; 3, $\bar{3}$, $3m$, 32, $\bar{3}\ 2m$	One threefold rotation axis	Rhombohedral (trigonal)
E	14–20; 4, $\bar{4}$, $4/m$, 422, $4mm$, $4/m2/m2/m$, $\bar{4}2m$	One fourfold rotation axis	Tetrgonal
F	21–27; 6, $\bar{6}$, $6/m$, $6mm$, 622, $\bar{6}2m$, $6/m\ 2/m\ 2/m$	One sixfold rotation axis	Hexagonal
G	28–32; 23, $2/m\ \bar{3}$, $\bar{4}\ 3m$, 432, $4/m\ \bar{3}\ 2/m$	Four threefold rotation axes at 70° 32/to each other	Cubic

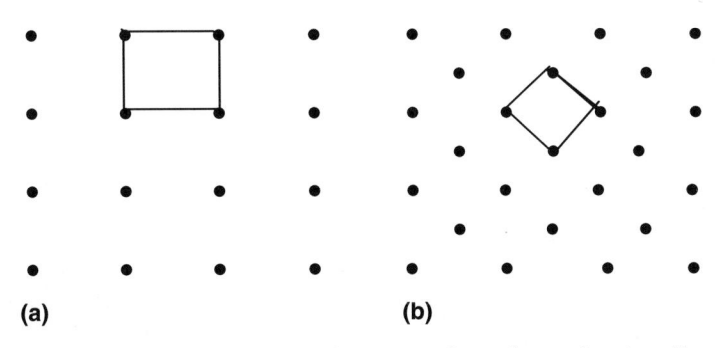

Fig. 4.2. Different pattern but same shape (*square*) unit cell

These seven sets into which the 32 point groups can be regrouped are known as *seven crystal systems*. There is one common symmetry element between the point groups belonging to each set, that is, crystal systems and they are named as expressed in Table 4.2.

4.3 Crystal Systems

These crystal systems some times are also named as crystal classes. Now as the unit cells (Fig. 4.2) are the building block of the two three-dimensional patterns, that is, space lattices, they also show the same symmetry as that of the space lattices and they are characterized by relations between the axial lengths known as unit translational vectors and the angles between them. It is more convenient to express these edge distances of the unit cells as vectors and so they will henceforth be noted as unit translational vectors: **a, b**, and **c** and the angles between them as α, β, and γ. Table 4.3 lists those unit cells and the relations between the unit translational vectors.

Table 4.3. Unit cells of seven crystal systems

Set	Crystal class or systems	Relation between a, b and c	Relation between α, β and γ
A	Triclinic	$a \neq b \neq c$	$\alpha \neq \beta \neq \gamma$
B	Monoclinic	$a \neq b \neq c$	$\alpha = \gamma = 90°° \neq \beta$
C	Orthorhombic	$a \neq b \neq c$	$\alpha = \beta = \gamma = 90°$
D	Rhombohedral	$a = b = c$	$\alpha = \beta = \gamma \neq 90°$
E	Tetragonal	$a = b \neq c$	$\alpha = \beta = \gamma = 90°$
F	Hexagonal	$a = b \neq c$	$\alpha = \beta = 90°, \ \gamma = 120°$
G	Cubic	$a = b = c$	$\alpha = \beta = \gamma = 90°$

Note: These seven unit cells represent the total number of different ways in which it is possible to draw a parallelepiped. However, it should be mentioned here that the sign $\neq$, which means 'not equal to' in mathematics, means here usually not equal to by reason of symmetry and accidentally the equality may occur

4.4 Bravais Lattices

Now a question may be raised: "Is it possible to create different patterns two-dimensional or three-dimensional and yet to preserve the same symmetry present in the unit cell, that is, keeping the shape of the unit cells unchanged?"

The answer to this question is "yes." Let us see how this is possible from an example of plane square lattice.

Like this example in plane lattice, in space lattice also by assigning additional lattice points in the unit cells it is possible to create a new three-dimensional pattern but yet keeping the shape of the unit cells unchanged. This assignment of additional lattice points may be done by several possible ways and these result in unit cells as follows:

1. Primitive with only corner lattice points   Symbol P
2. Base-centered, i.e., a face-centered   Symbol A
   Or B face-centered   Symbol B
   Or C face-centered   Symbol C
   Or all face-centered   Symbol F
3. Body-centered   Symbol I

Now as there are seven numbers of crystal systems and there are six numbers of different lattices that could be constructed by assigning additional lattice points like P, A, B, C, F, and I, so it might be possible to find 42 (6×7) different types of space lattices. But we get actually only 14 numbers distinctly different lattices known as *Bravais lattices*.

Any additional lattice other than these 14 in number is a repetition of any one of the existing lattices. Now, those possible 14 numbers of Bravais lattices distributed in seven numbers of crystal systems and those which are redundant are given in Table 4.4.

Table 4.4. The possible Bravais lattices

Set	Crystal class or system	Symbol of possible space lattices	Total no. possible in the class	Total no. of space (Bravais) lattices
A	Triclinic	P	1	1
B	Monoclinic	P and C	2	$2 + 1 = 3$
C	Orthorhombic	P,C,I, and F	4	$3 + 4 = 7$
D	Rhombohedral	P	1	$7 + 1 = 8$
E	Tetragonal	P and I	2	$8 + 2 = 10$
F	Hexagonal	P	1	$10 + 1 = 11$
G	Cubic	P,I, and F	3	$11 + 3 = 14$

Table 4.5. The redundant space lattices

Set	Crystal class or system	Symbol of redundant space lattices	Total no. of redundant lattice in the class	Total no. of redundant space lattices	Reason for redundancy
A	Triclinic	A,B,C,I, and F	5	5	All others are repetitions
B	Monoclinic	A,B,I, and F	4	$5 + 4 = 9$	All others are repetitions
C	Orthorhombic	A and B	2	$9 + 2 = 11$	All others are repetitions
D	Rhombohedral	A,B,C,I, and F	5	$11 + 5 = 16$	All others are repetitions
E	Tetragonal	A,B,C, and F	4	$16 + 4 = 20$	All others are repetitions
F	Hexagonal	A,B,C,I, and F	5	$20 + 5 = 25$	All others are repetitions
G	Cubic	A,B, and C	3	$25 + 3 = 28$	All others are repetitions

The possible Bravais lattice depends on the fact by the addition of extra lattice sites in the crystal pattern, the pattern changes without changing the basic primitive characteristics [1–3]. A body-centered cubic is a changed pattern but lattice represents basically a cubic lattice. It may result in the same lattice but does not repeat with other lattices (Table. 4.5). Figure 4.3 demonstrates this fact. Some examples of the redundant space lattices are shown in Fig. 4.4a–h.

Conclusion: Therefore, all crystals have a space lattice that must be one of the 14 Bravais lattices and not more, as for these are the only ways in which indistinguishable points can occur uniquely in three dimensions of space.

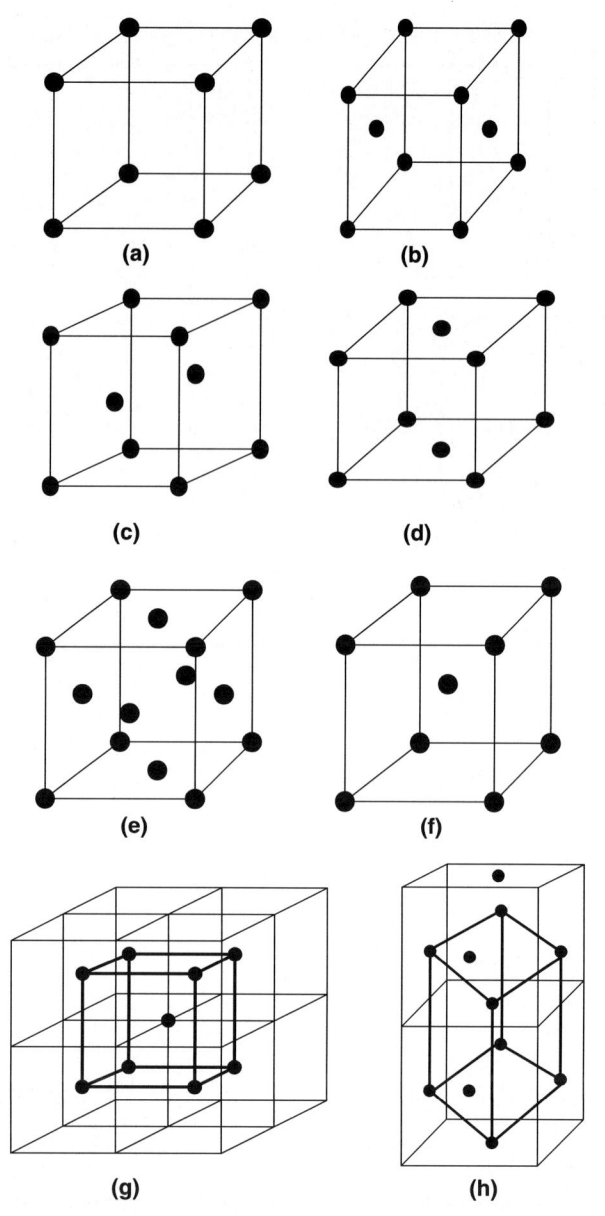

Fig. 4.3. (a) Corner sites, primitive symbol: P. **(b)** A face-centered, symbol: A. **(c)** B face–centered, symbol: B. **(d)** C face–centered, symbol: C. **(e)** All face-centered, symbol: F. **(f)** Body–centered, symbol: I. **(g)** A possible body-centered cubic cell basically remains cubic (*bold*) $a = b = c$ and $\alpha = \beta = \gamma = 90°$, yet the pattern is changed and there is no repetition except remaining as body-centered cubic. All the corner atoms are not shown. **(h)** A possible face-centered cubic cell which basically remains cubic (*bold*) even after changing the pattern. $a = b = c = a$ of the original primitive. The corner atoms are not shown

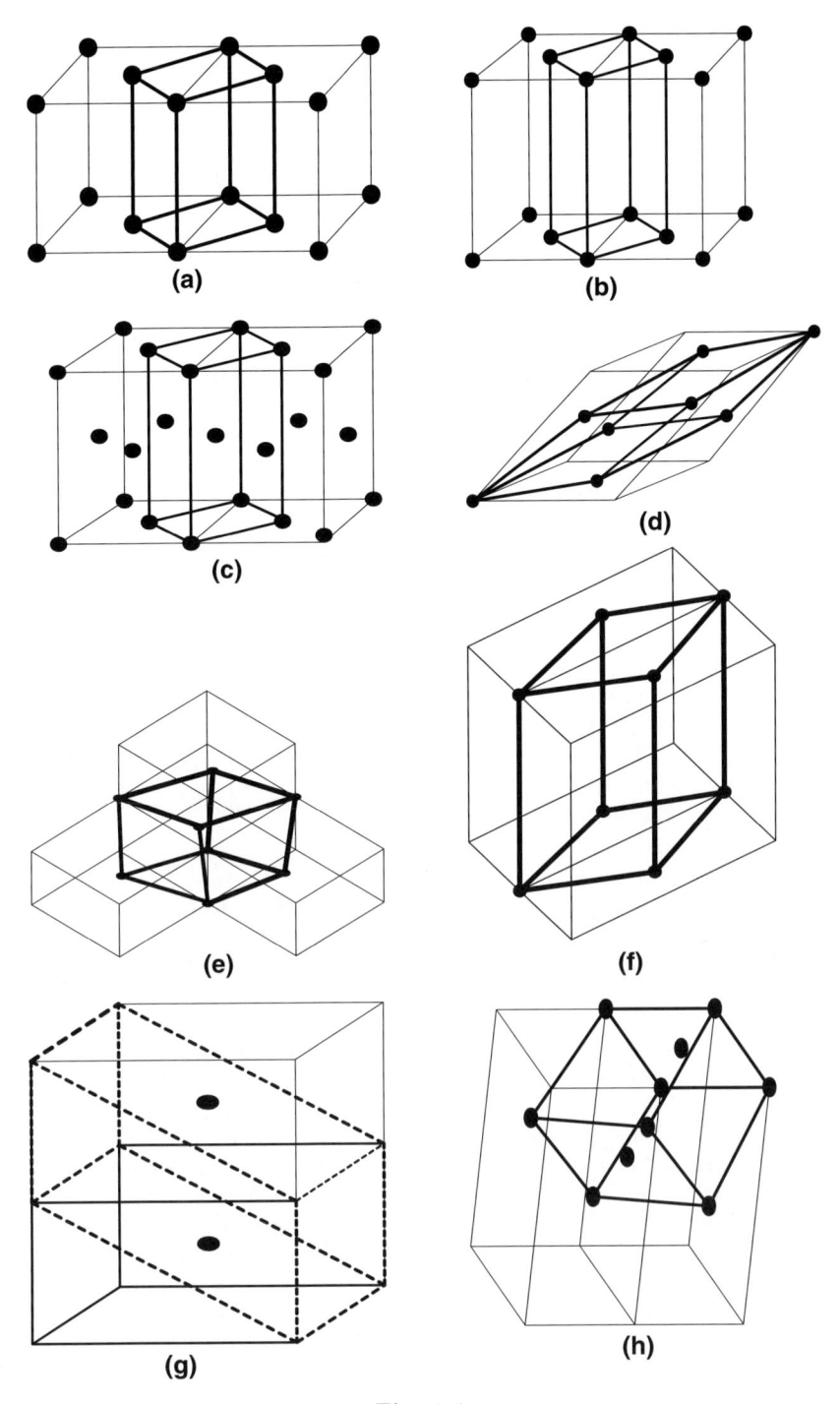

Fig. 4.4.

Some important materials as an example of 14 Bravais lattices they belong to are as follows:

Set	Crystal system	Bravais lattice	Example
A	Triclinic	Primitive	Copper sulfate, sodium bisulfate
B	Monoclinic	Primitive	Malachite, azunite
		Base centered	Gypsum, mica
C	Orthorhombic	Primitive	Topaz, aragonite
		Base centered	Chlorine, iodine
		Body centered	Thorium tetrafluoride
		Face centered	Sulfur, plutonium
D	Rhombohedral	Primitive	Quartz, tourmaline, antimony, arsenic
E	Hexagonal	Primitive	Zinc, titanium, magnesium, cadmium
F	Tetragonal	Primitive	Rutile, cassiterite
		Body centered	Tin, indium
G	Cubic	Primitive	Polonium, sodium chlorate
		Body centered	Iron, chromium, sodium, potassium, tungsten, vanadium
		Face centered	Copper, silver, gold, diamond, sodium chloride

The number of lattice points, N (atoms or molecules in actual crystals), in a Unit Cell is given by $N = 1 + (1/2)f + b$, where f and b stand for number of points in the centre of the faces and at the centre of the body of the unit cell.

Fig. 4.4. (Continued) (a) The addition of base-centered points in cubic cell results in primitive tetragonal space lattice. The resultant lattice is shown by *heavy lines*. $a = b \neq c$ and $\alpha = \beta = \gamma = 90°$. The resultant tetragonal lattice is shown by *bold lines*. (b) The addition of base-centered points in primitive tetragonal cell results in primitive tetragonal space lattice and does not yield to a new lattice. (c) The addition of face-centered lattice points in tetragonal lattice results in body-centered tetragonal space lattice points and thus does not give a new lattice and resultant lattice, that is, body-centered tetragonal (*shown by heavy lines*) is already considered as a new lattice. (d) The face-centered lattice in Rhombohedra system is equivalent to primitive rhombohedra lattice. The corner points are not shown and the resultant lattice is shown by *heavy lines*. (e) The figure shows the equivalence between body-centered rhombohedra and the primitive rhombohedra lattice. The resultant lattice is shown by heavy lines. (f) The figure shows the equivalence between base-centered and primitive (*shown by heavy lines*) lattices in monoclinic system. No corner atoms are shown. (g) & (h) The equivalence between bcc (g) and fcc (h) monoclinic with C (base) centered monoclinic. This shows that only primitive and C face-centered monoclinic lattices are possible, rest are repetitions and redundant. No corner atoms are shown

A primitive cubic lattice unit cell has atoms at the corners but each one of them is shared by eight neighboring unit cells and therefore the total contribution of corner atoms is equivalent to only one. A body-centered cubic unit cell has only two ($f = 0$) and a face-centered cubic unit cell has four, one due to eight corner points and three due to centre points on each of six faces. These face centered points are shared by two neighboring unit cells.

4.5 Summary

1. Crystal symmetry considers the scheme of positional repetitions as well as the influence of motifs on the symmetry.
2. When the existence of the motifs is only considered on planes in three-dimensional crystal, the different patterns that can be created are called "point groups".
3. There can be only 32 of such point groups as more than that will be a duplication of the one that already exists.
4. These 32 point groups can be grouped into seven different classes, each having different shapes of the unit cells.
5. A creation of additional lattice points without disturbing the class makes 14 different types of patterns, which appears different as a whole but basically remains in the same class and they are known as Bravais lattices.

References

1. L.V. Azaroff, *Introduction to Solids* (Mc Graw Hill, New York, 1960)
2. M.J. Buerger, *X-ray Crystallography* (Wiley, New York, 1942)
3. C. Kittel, *Introduction to Solid State Physics*, 5th edn. (Wiley, New York, 1976)

5

Crystal Symmetry (Crystal Pattern): II

5.1 Microscopic Symmetry Elements in Crystals

So far we have discussed the macroscopic symmetry elements that are manifested by the external shape of the three-dimensional patterns, that is, crystals. They can be studied by investigating the symmetry present in the faces of the crystals. In addition to these symmetry elements there are two more symmetry elements that are related to the detailed arrangements of motifs (atoms or molecules in actual crystals). These symmetry elements are known as microscopic symmetry elements, as they can only be identified by the study of internal arrangement of the motifs. As X-ray or electron diffraction can reveal the internal structures, these symmetry arrangements can only be identified by X-ray, Electro or Neutron diffraction. Obviously, they are not revealed in the external shape of the pattern. These symmetry elements are classified as microscopic symmetry elements. There are two such types of symmetry elements: (i) glide plane of symmetry and (ii) screw axis of symmetry.

Glide plane of symmetry: It is a combination of reflection and translation of the motif. It is explained by Fig. 5.1. Figure 5.2 shows simple pattern of a glide plane.

The translation associated with the glide plane may be one of the following (Fig. 5.3):

(i) One half of one of the unit translational vectors, that is, **a, b,** and **c,** which define the unit cell
(ii) One half or one quarter of the face diagonal

Hermann–Mauguin symbol of glide plane of symmetry:

Reflection + translation **a**/2 symbol = a
Reflection + translation **b**/2 symbol = b
Reflection + translation **c**/2 symbol = c
Reflection + $1/2$ face diagonal symbol = n
Reflection + $1/4$ face diagonal symbol = d

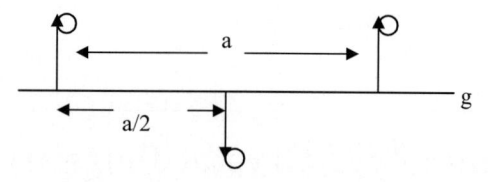

Fig. 5.1. a is the unit translational vector. The motif suffers a reflection on the mirror plane and undergoes a translation half the way. g the mirror perpendicular to the diagram and is known as the glide plane

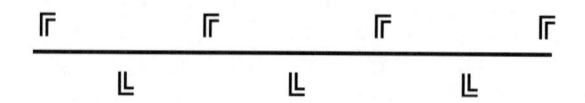

Fig. 5.2. A border showing simple glide plane of symmetry

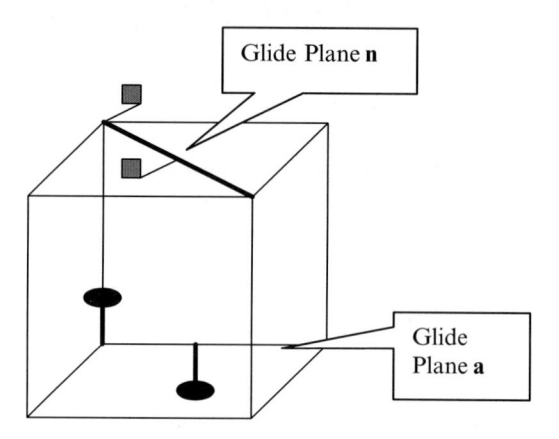

Fig. 5.3. Reflection and $1/2$ translation along a, "a" glide, and along face diagonal, "n" glide

There are plenty of natural examples: The stacking sequence of the close-packed planes, that is, {111} sets of planes in face-centered lattice (fcc) are known as ABC.... The entire scenario of stacking of these close-packed planes can be viewed as the stacking of hard billiard balls one over the other (Fig. 5.4). It can be easily visualized that a layer of such balls can never be placed exactly vertically at the top as the arrangement has to get toppled. To make a stable arrangement, the balls of one layer have to be placed on the gaps of the balls in the lower layer. This can be understood from the diagram as below.

Now one particular ball will have glide plane of symmetry on the same plane the ball exists, but will also have a glide plane up the plane in three-dimensional arrangements of the balls. The different layers arranged one over other are manifested by different colors.

Stacking sequence of Billiard balls

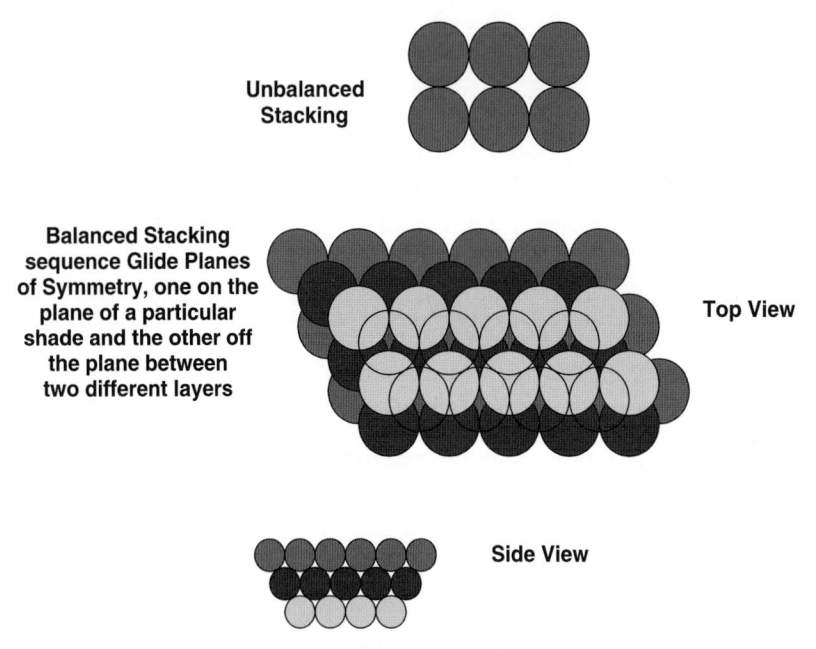

Fig. 5.4. The balanced stacking sequence of Billiard balls: *red* is the first layer, *blue* the second, and *green* the third. The fourth layer shown by open circle is the repetition of the first layer, that is, it can stay exactly over first layer. Each layer of first to third has suffered a glide in the plane over them

Note: A glide plane is a plane across which mirror reflection combined with a translation transforms an array of motifs into self coincidence. This stacking sequence shown above for Billiard balls has an important resemblance with that of stacking sequence of crystal planes discussed later.

Screw Axis of Symmetry: It is the combined effect of rotation and translation, which transforms the array of motifs into self coincidence. The rotation axis is known as screw axis of rotational symmetry or simply screw axis of symmetry (Figs. 5.5 and 5.6).

Note: A screw axis of symmetry is an axis about which a rotation combined with translation parallel to the axis transforms an array of motifs into self coincidence.

All the rotation axis of symmetries obtained in macroscopic symmetry elements are also valid here only with the addition of a translation along the axis about which the rotation takes place. The translation is a fraction of the unit translational distance.

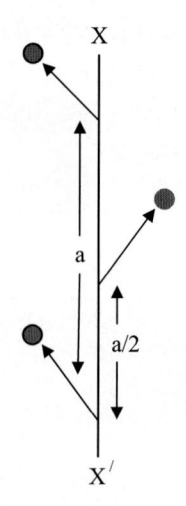

Fig. 5.5. A twofold rotation and half translation $a/2$ along the rotation axis XX; Symbol: 2_1

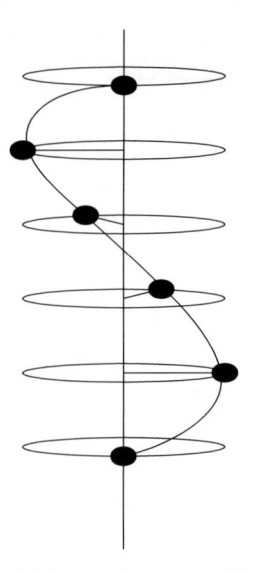

Fig. 5.6. A right handed screw in three dimension

Like this, if we consider all the possible rotation axes of symmetries and the possible translations along the rotation axis, henceforth known as screw axis, we get eleven different types of Screw axes of symmetries. They are listed in Table 5.1.

Conclusion: The microscopic symmetry elements are those symmetry elements that have no influence on the external shape of freely grown crystals, being concerned only with the detailed array of motifs (atoms or molecules)

Table 5.1. Eleven different types of screw axes of symmetries

Rotation axis	Translation: fraction of the distance between lattice points in the direction of the axis	Symbol (Hermann–Mauguin)
2	1/2	2_1
3	1/3	3_1
3	2/3	3_2
4	1/4	4_1
4	1/2	4_2
4	3/4	4_3
6	1/6	6_1
6	1/3	6_2
6	1/2	6_3
6	2/3	6_4
6	5/6	6_5

within the three-dimensional patterns (crystals) and involving a translation operation. They are detectable only by X-ray diffraction methods and hence the name "microscopic symmetry." They can only be manifested in space and so they are called spatial symmetry. These elements consist of FIVE different kinds of GLIDE planes and ELEVEN different types of SCREW AXIS.

It now can be appreciated those only symmetry elements that might be present in the structure of any crystal which is nothing but a three-dimensional pattern are the following:

Macroscopic symmetry   Mirror plane m
Centre of symmetry $\bar{1}$
Rotation axis 1, 2, 3, 4, 6
Rotary inversion $\bar{3}, \bar{4}, \bar{6}$

Microscopic symmetry   Glide plane a, b, c, n, d
Screw axis $2_1, 3_1, 3_2, 4_1, 4_2, 4_3$
$6_1, 6_2, 6_3, 6_4, 6_5$

The detailed information about the symmetry properties of the complete array of atoms or molecules in a crystal can be obtained only from the combined specifications of the symmetry at a lattice point, that is, the point groups modified by the microscopic translational symmetry and the distribution in space (Bravais lattice) of those points.

Such combination demonstrates the full description of the symmetry elements present in any crystal and it is named as *space group* [1, 2].

Conclusion: The *space group* of a crystal is the collection of symmetry elements (macroscopic and microscopic) which, when considered to be distributed in space according to the Bravais Lattice, provides knowledge of total symmetry present in the crystal amongst the different array of atoms or molecules within it.

5.2 Space Groups

Therefore, a space group is a possible combination of all the symmetry elements, macroscopic and microscopic, in space of the Bravais lattice and can be derived. It is found that when all such symmetry elements are combined and applied in the Bravais lattices, 230 different types of crystal space lattices are possible. It is appropriate to mention here that any crystal either naturally free grown or crystallized artificially from the solutions of the synthesized compounds must belong to any of these possible 230 types of space groups [1,2].

Note: The Hermann–Mauguin space group notation for any particular crystal comprises two parts. The first part identifies the Bravais lattice type into which the crystal belongs and the second part identifies the total symmetry of the array of atoms in the crystal and therefore also the crystal system. In the second part that identifies the symmetry, only those symmetry elements are included in the symbol that are necessary to describe the space group uniquely. The remainders are being omitted since they follow, as a necessary consequence.

Example:
P m m m = Primitive lattice, three mirror planes at right angles.

   P n m a = Primitive lattice, glide plane (n), that is, $1/2$ diagonal face ($b/2 + c/2$) and perpendicular to **a** axis, mirror plane perpendicular to **b** axis, and finally another glide plane (a) perpendicular to **c** axis with translation $a/2$. The same space group oriented differently might be P n a m, P b n m, P c m n, P m n b, and P m c n.

   F m 3 m = Face centered, mirror plane along a, threefold rotation axis along b, and another mirror plane along c axis.

   It is instructive to derive a few space groups as follows:

Triclinic System: There is possibility of the existence of only one primitive P lattice. It has two point groups 1 and 1-bar, that is, onefold rotation and a centre of symmetry and these are the only ways in which the symmetry elements can occur in triclinic system. These two combinations, i.e., P 1 and P $\bar{1}$ are two space groups of the triclinic system.

Monoclinic System: It has two Bravais lattices, i.e., primitive (P) and base-centered C, and three point groups 2, m, and $2/m$. In detailed study of symmetry, the array of atoms that constitutes the structure of the crystal, a macroscopic mirror plane m, might be a glide plane c, while twofold rotation axis might be a screw axis as 2_1. Considering these aspects of possible symmetry, the complete set is given as follows:

1. A twofold rotation axis in P and C lattices = P2, C2
2. A 2_1 screw axis in P and C lattices = P2_1, C2_1
3. A mirror plane in P and C lattices = Pm, Cm
4. A glide plane c in P and C lattices = Pc, Cc

5. A twofold axis and mirror plane in P and C $= $ P2/m, C2/m
6. A 2_1 screw axis and mirror plane in P and C $=$ P2$_1$/m, C2$_1$/m
7. A twofold axis and a glide plane c in P and C $=$ P2/c, C2/c
8. A 2_1 screw axis and glide plane c in P and C $=$ P2$_1$/c, C2$_1$/c

However, out of these 16 numbers of combinations as mentioned above do not specify different arrays of symmetry elements in space, as C2 and C2$_1$ are identical and combinations C2$_1$/c and C2/c are also identical. So, there can be only thirteen (13) different combinations of symmetry elements that can occur in the array of atoms in a monoclinic system and so the monoclinic system has 13 space groups.

Note: When all possible but different combinations of symmetry elements present between array of atoms in seven crystal systems are added together, we get 230 number of space groups (Table 5.2).

These 230 space groups are the only ways in which different distribution of compatible combinations of macroscopic and microscopic symmetry elements can occur in the array of atoms in any crystal.

In reading a space group notation it is important to remember the following:

A onefold axis includes the elements 1, $\bar{1}$
A twofold axis includes the elements 2, 2_1, m, a, b, c, n, d
A threefold axis includes the elements 3, $\bar{3}$, 3_1, 3_2
A fourfold axis includes the elements 4, $\bar{4}$, 4_1, 4_2, 4_3
A sixfold axis includes the elements 6, $\bar{6}$, 6_1, 6_2, 6_3, 6_4, 6_5;
these may further be noted that the existence of
onefold axes denote the triclinic system,
one twofold axis denotes the monoclinic system,
three twofold axes at right angles denote the orthorhombic system,
one threefold axis denotes the rhombohedra system,
four threefold axes at 70° 32/to one another denote the cubic system,
one fourfold axis denotes the tetragonal system, and
one sixfold axis denotes the hexagonal system.

Table 5.2. Possible space groups

System	No of space group
Triclinic	2
Monoclinic	13
Orthorhombic	59
Rhombohedral	25
Hexagonal	27
Tetragonal	68
Cubic	36
Total	230

5.3 Constitution of Space Groups

The detailed study to know how these space groups are constituted from seven crystal systems is out of the scope of this book. Mathematically it can be derived from the group theory and geometrically, the derivation of all these 230 space groups is cumbersome and therefore, for only one crystal system, that is, for monoclinic system the procedure that is adopted is explained below.

A particular crystal system has some definite number of point groups and for this monoclinic system it has symmetry operations like 2, m, and $2/m$, that is, twofold rotation, a mirror plane, and twofold with mirror plane of symmetries. Now, for three-dimensional crystal the possible symmetry elements will include also screw axes and glide planes, and when screw axes and glide planes are added to the point group of symmetries for this system, we can say that different possibilities that may exist are 2, 2_1, m, c, $2/m$, $2_1/m$, $2/c$, and $2_1/c$. Now each of these symmetry groups are repeated by lattice translation of the Bravais lattices of that system. As monoclinic system has only primitive P and C, all the symmetry possibilities may be associated with both P and C. Therefore, if they are worked out, they come out to be 13 in number and they are Pm, Pc, Cm, Cc, P2, P2_1, C2, P$2/m$, P$2_1/m$, C$2/m$, P$2/c$, P$2_1/c$, C$2/c$, etc.

Similarly the space groups of all the crystal systems can be worked out and these come out to be as mentioned earlier 230 in number.

5.4 Summary

1. When two more symmetry operations move the motifs out from their original plane on which they originally exist, more lattice patterns are created.
2. These symmetry operations are called microscopic symmetry operations as they can only be identified from internal structure of the crystal lattice in three dimensions and not by the geometrical shape of the crystal.
3. The number of possible patterns comprising the lattice sites not confined on a plane will then increase to 230 and these lattices are called "space groups."

References

1. M.J. Buerger, *X-ray Crystallography* (Wiley, New York, 1966)
2. M.M. Woolfson, *An Introduction to X-ray Crystallography* (Cambridge University Press, Cambridge, 1970)

6

Crystals and X-Ray

X-ray was discovered by the German Physicist Roentgen in 1895 almost accidentally while performing experiments on the discharge of electricity through gases under low pressure. Another German Physicist Von Laue's discovery of regular diffraction pattern when X-ray was diffracted by single crystals opened a new year of crystal structure analysis. It is only after the Von Laue's famous experiment; it was proved that the crystallinity of a solid depends on the regular arrangement of the constituent atoms or molecules in three-dimensional space and not on the external features. It should also be known that the wavelength of X-ray radiation varies between ~ 1 and ~ 3 Å, which is the order of interatomic distances of the solids and that is why X-ray satisfies the diffraction conditions. In addition to that of X-ray, there are two other methods of analysis of crystallinity of materials and they are electron and neutron diffraction. These three techniques have their own advantages and limitations and in essence complement one another.

6.1 Production and Properties of X-Ray

X-rays are produced when accelerated electrons while penetrating through the target material and moving through the orbital electron cloud of atoms of the target are rapidly decelerated by the resistive force. Due to this deceleration of the bombarding electrons, their energy is emitted in the form of X-ray. This spectrum of X-ray radiation continuously varies in intensity from a lowest wavelength value known as *short-wavelength limit*. This radiation spectrum is named as continuous or general radiation. When the accelerating potential for the bombarding electrons is increased, then these bombarding even after being decelerated still possess sufficient energy to knock off electrons from different orbits of the target atoms. When K-shell electrons are knocked off, then the target atoms are raised up to a potential energy corresponding to the K-shell. The electrons from the higher energy levels like L-shell or M-shell then jump down to the K-shell to fill up the vacancy and lower down the potential energy

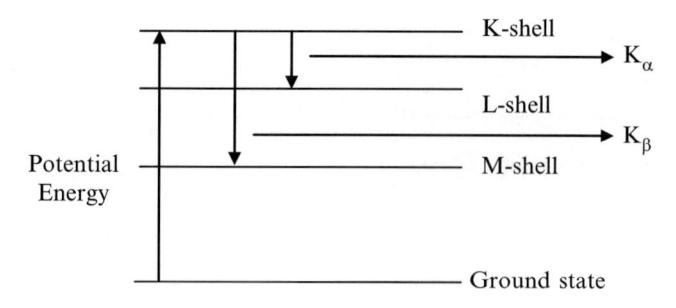

Fig. 6.1. Emission of K_α and K_β

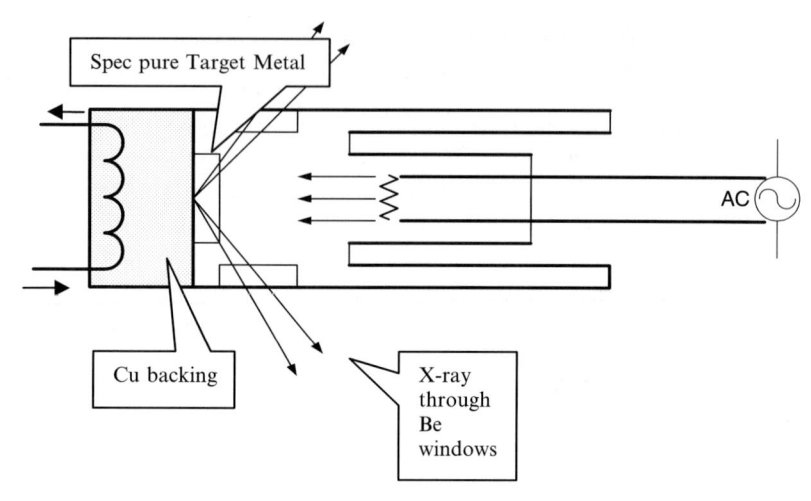

Fig. 6.2. X-ray tube (sealed type) showing target metal, cooling device, and the filament for electron emission

of the target atoms. These processes emit radiation of specific wavelengths which are characteristic of the target atom as they depend on the atomic number (Z) of the atoms of the target. These radiations being dependent on the atomic number are named as *characteristic radiation* [1, 2].

In Figs. 6.1, 6.2, and 6.3, potential energy change and subsequent emissions of K characteristic radiations, the X-ray tube used for the production of X-rays, and the X-ray spectra showing both the continuous and characteristic radiations are shown, respectively. Now, whenever X-rays are emitted from the metal target, the radiation contains both the general and characteristic radiations, and in most of the diffraction procedures monochromatic radiation is required. This monochromatization of X-ray spectra is obtained by absorbing the K_β radiation by an absorber (filter) of optimum thickness. Obviously, different targets must have different K_β filters. The transmitted radiation through this absorber contains only K_α radiation and the background (the part of general radiation). Now, to eliminate the background from transmitted

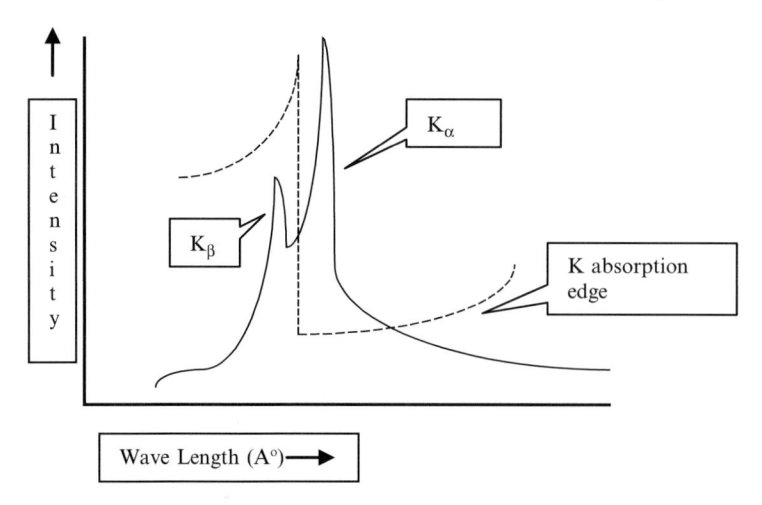

Fig. 6.3. X-ray spectra from Cu target and the position of K absorption edge of Ni used as K_β filter for Cu radiation

radiation, balanced filters (Ross filters) or crystal monochromators are used. The crystal monochromator and absorption edges of these balanced filters are shown in Fig. 6.4a, b.

6.2 Laue Equations

It has been stated before that the Laue diffraction pattern from single crystals was the beginning of the idea that the internal regular arrangement of atoms in space is responsible for the crystallinity of the material. Figure 6.5 shows the X-ray diffraction from a row of atoms.

Now, when the path difference between incident and diffracted beams becomes equal to the integral multiple of the wavelength λ, then the interference maxima condition will be satisfied, i.e., $BC - AD = a\cos\theta - a\cos\varphi = n\lambda$, this is one-dimensional Laue equation. When the other two directions are considered, then the corresponding Laue equations are

$$b\,\cos\theta' - b\,\cos\varphi' = n\lambda \quad \text{and} \quad c\,\cos\theta'' - c\,\cos\varphi'' = n\lambda.$$

In vector form, these three Laue equations can be written as

$$\mathbf{a}\cdot(\mathbf{S} - \mathbf{S}_0) = n\lambda, \quad \mathbf{b}\cdot(\mathbf{S} - \mathbf{S}_0) = n\lambda, \quad \text{and} \quad \mathbf{c}\cdot(\mathbf{S} - \mathbf{S}_0) = n\lambda.$$

When these three conditions are simultaneously satisfied, the entire diffraction phenomenon may be equivalent to a planer reflection and gives rise to Bragg's law. It will be shown in this chapter that the wavelengths for which the Laue conditions are satisfied are not the characteristic radiation but general radiation.

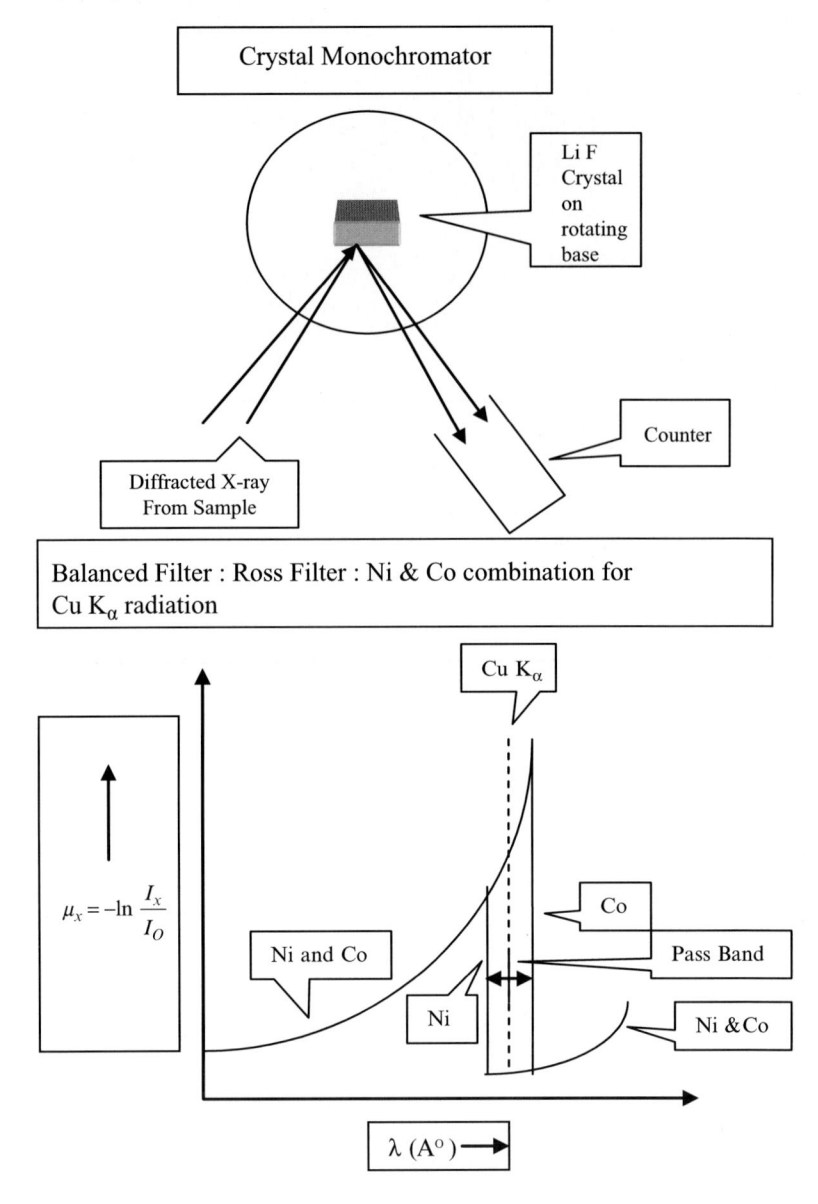

Fig. 6.4. (a) Crystal monochromator. LiF single crystals can diffract like other crystals, K_α and K_β radiations from any target at specific angles and the counter placed at such angles will only record the diffraction pattern due to that wavelength and that without reducing the intensity of the radiation and reducing considerably the background. This is the major advantage of crystal monochromator now attached invariably with diffractometer. (b) Ross filter. As mass absorption coefficient is dependent on λ^3, the two absorbers differing in Z by one will give same absorption coefficient for any wavelength except for the narrow region called *pass band*. This is the narrow region between K absorption edges of these two absorbers and when this pass band is chosen to include the K_α, the result is a strong monochromatic beam

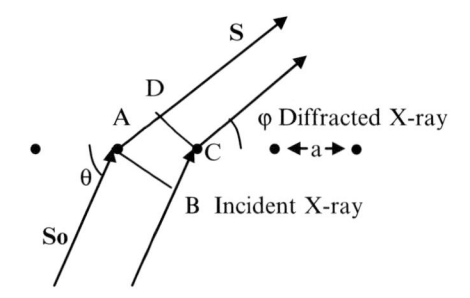

Fig. 6.5. X-ray diffraction from the rows of atoms/molecules

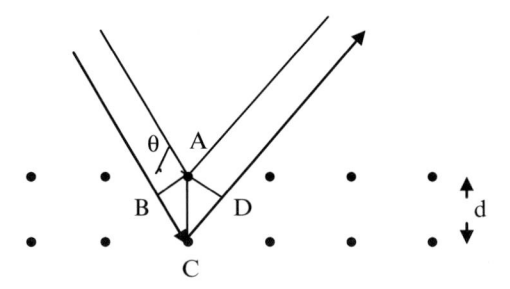

Fig. 6.6. Bragg's reflection from a set of hkl planes having interplanar distance d

6.3 Bragg's Law

Let us consider two parallel crystallographic planes defined by the Miller indices hkl having their planes perpendicular to the diagram (Fig. 6.6) and X-ray having definite wavelength, λ is incident at a particular angle.

The path difference between the incident and diffracted wavefronts AB and AD will be $BC - CD = 2d\sin\theta$ and when this is equal to an integral multiple of the characteristic wavelength λ, then there will be "reflection" and the reflected beams will interfere constructively, i.e., $2d\sin\theta = n\lambda$. This is known as *Bragg's law* and the angle θ is known as *Bragg angle*. The d is the interplanar spacing. As the values of the interplanar spacing, i.e., d are different for different sets of planes, the angle θ will also be dependant on the plane for a definite wavelength of X-ray radiation. So, the Bragg's condition for reflection can be more generalized as [3]

$$2d_{hkl}\sin\theta_{hkl} = n\lambda.$$

6.4 Reciprocal Lattice

The concept and the applicability of reciprocal lattice are very important in structural analysis of crystalline materials. The importance of this hypothetical lattice can be best understood as we move through the following chapters.

It can be rightly said that "The reciprocal lattice is as important in crystal structure analysis as the walking stick of a blind man moving in a narrow lane having frequent turns." It is extremely difficult if not impossible to picture the different intersecting crystal planes satisfying the Bragg's reflection in three-dimensional lattice from the two-dimensional array of spots or lines.

The problem was reduced by assuming normal on such planes and as the angles of these normal are equal to the angles between the intersecting planes. Such imaginary normal when projected on the surface of a sphere or a plane, the projections are, respectively, known as *spherical projections* or *stereographic projections*. But to minimize the problem of finding the angles and the interplanar distances between any two sets of planes, the introduction of the concept of a hypothetical lattice becomes essential. Let us introduce the concept from an optical analogy. When a distant light source is observed through a fine mesh, e.g., when the Sun is viewed through a silk umbrella, a diffraction pattern is observed. This is because the distances between two fine fibers of silk cloth is approximately of the order of wave length of light. It can be seen from the diffraction pattern that the distance of separation between the central spot and the second spot either on the x-axis or on the y-axis is found to be twice the distance between the central spot and the first spot. The entire diffraction phenomenon is shown in Fig. 6.7. The beautifully arranged diffraction spots form a two-dimensional lattice and can be utilized to introduce the reciprocal lattice. If the three unit translational vectors which define the direct lattice are given as $\mathbf{a}$, $\mathbf{b}$, and $\mathbf{c}$, then corresponding to these vectors let us introduce three other vectors such as $\mathbf{a}^*$, $\mathbf{b}^*$, and $\mathbf{c}^*$, so that the following relations hold good:

$$\mathbf{a}\cdot\mathbf{a}^* = 1, \quad \mathbf{b}\cdot\mathbf{b}^* = 1, \quad \text{and} \quad \mathbf{c}\cdot\mathbf{c}^* = 1 \quad \text{and also} \quad \mathbf{b}\cdot\mathbf{a}^* = \mathbf{c}\cdot\mathbf{a}^* = 0. \quad (6.1)$$

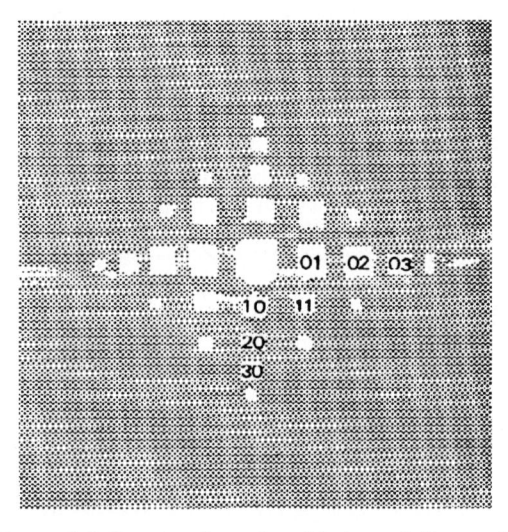

Fig. 6.7. Optical diffraction from fine fabric and distant light source

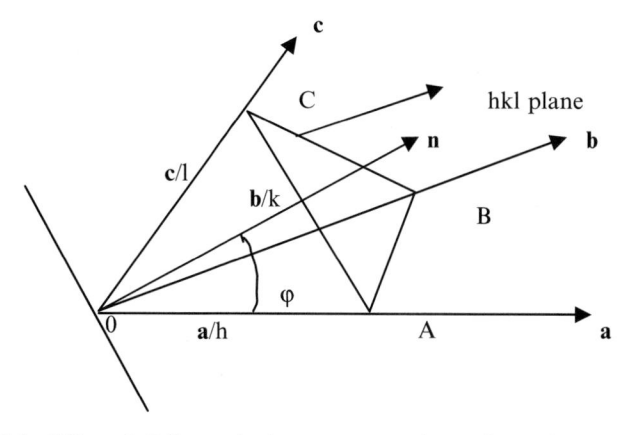

Fig. 6.8. *OA, OB,* and *OC* are the intercepts on the **a**, **b**, and **c** axes of the plane *hkl*, **n** is the unit vector normal to the plane *hkl* and φ is the angle that this unit vector makes with the **a**-axis

Now, as the dot product of $\mathbf{a}^*$ with $\mathbf{b}$ and $\mathbf{c}$ vanishes, therefore, $\mathbf{a}^*$ is perpendicular to both of them, i.e., perpendicular to the plane containing $\mathbf{b}$ and $\mathbf{c}$ in the direct lattice. This gives the direction of $\mathbf{a}^*$ and from the relation $\mathbf{a} \cdot \mathbf{a}^* = 1$, we can have the magnitude of it. Similarly, $\mathbf{b}^*$ is perpendicular to the plane containing $\mathbf{c}$, and $\mathbf{a}$ and $\mathbf{c}^*$ are perpendicular to the plane containing $\mathbf{a}$ and $\mathbf{b}$. So, as $\mathbf{a}$, $\mathbf{b}$, and $\mathbf{c}$ the three vectors define the actual lattice (called "direct lattice"), the three vectors introduced as $\mathbf{a}^*$, $\mathbf{b}^*$, and $\mathbf{c}^*$ will also define another lattice. The relations between these vectors can be derived as [4]

$$\mathbf{a}^* = \frac{\mathbf{b} \times \mathbf{c}}{\mathbf{a} \cdot \mathbf{b} \times \mathbf{c}}, \quad \mathbf{b}^* = \frac{\mathbf{c} \times \mathbf{a}}{\mathbf{a} \cdot \mathbf{b} \times \mathbf{c}}, \quad \text{and} \quad \mathbf{c}^* = \frac{\mathbf{a} \times \mathbf{b}}{\mathbf{a} \cdot \mathbf{b} \times \mathbf{c}}.$$

Now, as $\mathbf{a} \cdot \mathbf{b} \times \mathbf{c}$ is the volume, V of the unit cell of the direct lattice, the above relations can be written as

$$\mathbf{a}^* = \frac{\mathbf{b} \times \mathbf{c}}{V}, \quad \mathbf{b}^* = \frac{\mathbf{c} \times \mathbf{a}}{V}, \quad \text{and} \quad \mathbf{c}^* = \frac{\mathbf{a} \times \mathbf{b}}{V}. \tag{6.2}$$

In Fig. 6.8, the unit translational vectors of the direct lattice and a plane having Miller indices *hkl* are shown to derive the further relations between the plane and the vectors in the corresponding "new imaginary lattice."

Now, let us introduce a vector $\mathbf{H}_{hkl}$ in the imaginary lattice which is defined as

$$\mathbf{H}_{hkl} = h\mathbf{a}^* + k\mathbf{b}^* + l\mathbf{c}^*.$$

This is logical as h, k, and l are pure numbers defining the plane *hkl* in the direct lattice.

Now,

$$OB + BA = OA, \quad \text{i.e.,} \quad BA = OA - OB = \mathbf{a}/h - \mathbf{b}/k.$$

So,

$$BA \cdot \mathbf{H}_{hkl} = (\mathbf{a}/h - \mathbf{b}/k) \cdot \mathbf{H}_{hkl} = (\mathbf{a}/h - \mathbf{b}/k) \cdot (h\mathbf{a}^* + k\mathbf{b}^* + l\mathbf{c}^*).$$

Now, using the relation between $\mathbf{a}$, $\mathbf{b}$, and $\mathbf{c}$ and $\mathbf{a}^*$, $\mathbf{b}^*$, and $\mathbf{c}^*$ from (6.1), we get

$$(\mathbf{a}/h - \mathbf{b}/k) \cdot \mathbf{H}_{hkl} = 1 - 1 = 0, \quad \text{i.e.,} \quad BA \cdot \mathbf{H}_{hkl} = 0$$

and similarly, we can also show that

$$CB \cdot \mathbf{H}_{hkl} = (\mathbf{b}/k - \mathbf{c}/l) \cdot \mathbf{H}_{hkl} = 0.$$

Therefore, the vector $\mathbf{H}_{hkl}$ of the imaginary lattice is perpendicular to the plane hkl of the direct lattice. Now, the interplanar distance of this hkl (d_{hkl}) is the normal distance of the plane hkl from the similar plane at 0 and it is given as

$$
\begin{aligned}
d_{hkl} &= (|\mathbf{a}|/h) \cos \varphi = (|\mathbf{a}|/h) \cdot \mathbf{n} \\
&= (|\mathbf{a}|/h) \cdot \frac{h\mathbf{a}^* + k\mathbf{b}^* + l\mathbf{c}^*}{|\mathbf{H}_{hkl}|} \\
&= 1/|\mathbf{H}_{hkl}|.
\end{aligned}
\tag{6.3}
$$

The vector $\mathbf{H}_{hkl}$ of the imaginary lattice defined as before is reciprocal of the interplanar distance in direct lattice and hence this "new" imaginary lattice is named as:

Reciprocal lattice. The vector $\mathbf{H}_{hkl}$ in the reciprocal lattice is the position vector of the lattice point hkl in that reciprocal space. Therefore, the hkl which are the Miller indices of a plane in direct space lattice are the coordinates of a lattice point in reciprocal space. This reciprocal lattice has also a similarity with the optical diffraction pattern of a distant light source when viewed through fine cloth (Fig. 6.7) which has been explained before.

Conclusion. To minimize the difficulty in the visualization of different intersecting planes in the direct lattice, a hypothetical lattice called reciprocal lattice is introduced in which each lattice point represents a plane of the direct lattice and the distance of each point from the origin of such lattice is the reciprocal of the interplanar distance of the plane that it represents in direct lattice.

6.5 Geometry of X-Ray Diffraction: Use of Reciprocal Lattice

Now is the time to realize importance of reciprocal lattice through the interpretation of X-ray diffraction phenomena. Let us imagine a sphere having radius equal to the reciprocal of the wavelength of the incident X-ray on a crystal stationed on the incident ray. Let the X-ray beam comprises of only the strongest K_α radiation (Fig. 6.9).

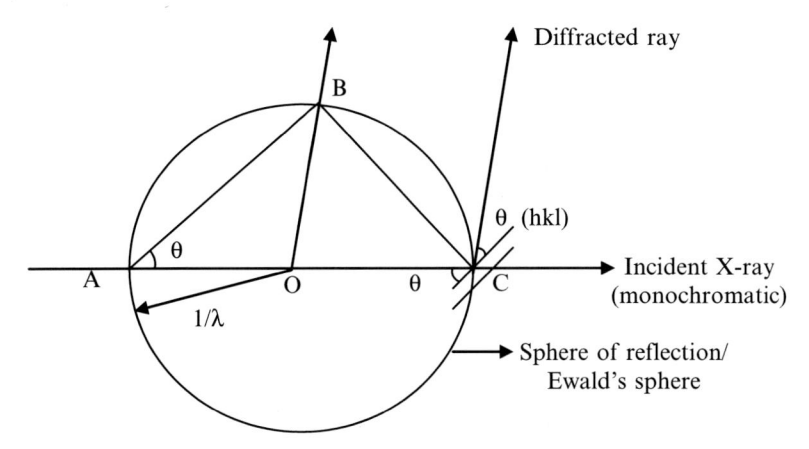

Fig. 6.9. Ewald's sphere of radius $1/\lambda$, crystal may be stationed either at the center O or at C. The reciprocal lattice origin is also at C, θ is the Bragg angle of reflection from the Miller plane (hkl) in the crystal, CB is drawn normal on the hkl plane

Now,

$$\angle ABC \colon \sin\theta = BC/AC = BC/(2/\lambda), \quad \text{i.e.,} \quad 2(1/BC)\sin\theta = \lambda. \quad (6.4)$$

But we have assumed that the X-ray has already satisfied the diffraction condition, i.e., Bragg's relation which is

$$2d_{hkl}\sin\theta = \lambda(n=1). \quad (6.5)$$

Therefore, to get (6.4) and (6.5) simultaneously satisfied, we get $BC = 1/d_{hkl}$, i.e., B is a point on the normal drawn on the plane hkl and its distance from the plane is reciprocal of the interplanar distance d_{hkl}. So, the point B is the reciprocal lattice point of the plane hkl, and when there is a Bragg's reflection from the plane for a definite wavelength then it should lie on the surface of the corresponding Ewald's sphere.

Conclusion. The Bragg's reflection will occur from a plane for a definite wavelength of X-ray, incident in a particular direction only when the reciprocal lattice point of the plane concerned lies on the surface of the corresponding Ewald's sphere.

It may be mentioned here that Laue spots (discussed in a later chapter) are obtained for general radiation and the spots vanish when a monochromator (K_β filter) is introduced before the beam. This is because out of all the reciprocal lattice points of the planes of the direct lattice only a very few or none at all will lie on the surface of the Ewald's sphere and satisfy the Bragg's condition, but when general radiation is used the wavelength increases from a short limit continuously and in that case plenty of the reciprocal lattice points may be assumed to lie on the surfaces of Ewald's spheres having continuously varying radius. This gives rise to plenty of Laue spots all symmetrically arranged as those of the reciprocal points in the reciprocal lattice.

The unit of the reciprocal lattice vectors $\mathbf{a}^*$, $\mathbf{b}^*$, and $\mathbf{c}^*$ is Å^{-1} or $\lambda\,\text{Å}^{-1}$ and in the latter case, it is unitless and they are some times expressed in reciprocal unit (r.u.). As α, β, and γ are the angles between the axes in the direct lattice, α^*, β^*, and γ^* will be the corresponding angles between the axes in the reciprocal space. It can be mentioned here that as in a particular type of lattice, different shapes of unit cells are possible (this has already been explained in an earlier chapter); in the corresponding reciprocal lattice, also different shapes of the unit cells are possible. However, like direct lattice, the reciprocal lattice remains same.

The relations between the unit translational vector of direct lattice and reciprocal lattice can be derived as follows.

The volume V of the unit cell in direct lattice is given by $V = \mathbf{a} \cdot \mathbf{b} \times \mathbf{c}$ and let $\mathbf{a}$, $\mathbf{b}$, and $\mathbf{c}$ are expressed as

$$\mathbf{a} = a_x\mathbf{i} + a_y\mathbf{j} + a_z\mathbf{k}, \quad \mathbf{b} = b_x\mathbf{i} + b_y\mathbf{j} + b_z\mathbf{k}, \quad \text{and} \quad \mathbf{c} = c_x\mathbf{i} + c_y\mathbf{j} + c_z\mathbf{k},$$

then the volume can be expressed in determinant form as

$$V = \begin{vmatrix} a_x & a_y & a_z \\ b_x & b_y & b_z \\ c_x & c_y & c_z \end{vmatrix}.$$

As the value of the determinant remains unchanged if the rows and columns are interchanged, we can write

$$V^2 = \begin{vmatrix} a_x & a_y & a_z \\ b_x & b_y & b_z \\ c_x & c_y & c_z \end{vmatrix} \times \begin{vmatrix} a_x & b_x & c_x \\ a_y & b_y & c_y \\ a_z & b_z & c_z \end{vmatrix}.$$

Applying the rule of multiplication of determinants,

$$V^2 = \begin{vmatrix} a_xa_x + a_ya_y + a_za_z & b_xa_x + b_ya_y + b_za_z & c_xa_x + c_ya_y + c_za_z \\ a_xb_x + a_yb_y + a_zb_z & b_xb_x + b_yb_y + b_zb_z & c_xb_x + c_yb_y + c_zb_z \\ a_xc_x + a_yc_y + a_zc_z & b_xc_x + b_yc_y + b_zc_z & c_xc_x + c_yc_y + c_zc_z \end{vmatrix}$$

or,

$$V^2 = \begin{vmatrix} \mathbf{a} \cdot \mathbf{a} & \mathbf{a} \cdot \mathbf{b} & \mathbf{a} \cdot \mathbf{c} \\ \mathbf{b} \cdot \mathbf{a} & \mathbf{b} \cdot \mathbf{b} & \mathbf{b} \cdot \mathbf{c} \\ \mathbf{c} \cdot \mathbf{a} & \mathbf{c} \cdot \mathbf{b} & \mathbf{c} \cdot \mathbf{c} \end{vmatrix}$$

using the vector dot product $\mathbf{i} \cdot \mathbf{i} = \mathbf{j} \cdot \mathbf{j} = \mathbf{k} \cdot \mathbf{k} = 1$ and $\mathbf{i} \cdot \mathbf{j} = \mathbf{i} \cdot \mathbf{k} = \mathbf{j} \cdot \mathbf{k} = 0$.

Now, $\mathbf{a} \cdot \mathbf{a} = a^2$, $\mathbf{b} \cdot \mathbf{b} = b^2$, $\mathbf{c} \cdot \mathbf{c} = c^2$, $\mathbf{a} \cdot \mathbf{b} = ab\,\cos\gamma$, $\mathbf{a} \cdot \mathbf{c} = ac\,\cos\beta$, and $\mathbf{b} \cdot \mathbf{c} = bc\,\cos\alpha$, then the above determinant is

$$V^2 = \begin{vmatrix} a^2 & ab\,\cos\gamma & ac\,\cos\beta \\ ab\,\cos\gamma & b^2 & bc\,\cos\alpha \\ ac\,\cos\beta & bc\,\cos\alpha & c^2 \end{vmatrix}$$

$$= a^2(b^2c^2 - b^2c^2\cos^2\alpha) + ab\,\cos\gamma(abc^2\cos\alpha\cos\beta - abc^2\cos\gamma)$$
$$+ ac\,\cos\beta(ab^2c\,\cos\gamma\cos\alpha - ab^2c\,\cos\beta)$$
$$= a^2b^2c^2(1 - \cos^2\alpha + 2\cos\alpha\cos\beta\cos\gamma - \cos^2\gamma - \cos^2\beta).$$

Therefore,

$$V = abc(1 - \cos^2 \alpha - \cos^2 \gamma - \cos^2 \beta + 2\cos\alpha \cos\beta \cos\gamma)^{1/2}.$$

Now as

$$V = \begin{vmatrix} a_x & b_x & c_x \\ a_y & b_y & c_y \\ a_z & b_z & c_z \end{vmatrix}, \qquad V^* = \begin{vmatrix} a_x^* & b_x^* & c_x^* \\ a_y^* & b_y^* & c_y^* \\ a_z^* & b_z^* & c_z^* \end{vmatrix},$$

where V^* is the volume of the unit cell of the reciprocal lattice

$$V \times V^* = \begin{vmatrix} 1 & 0 & 0 \\ 0 & 1 & 0 \\ 0 & 0 & 1 \end{vmatrix},$$

and so, $V = 1/V^*$.

Therefore, V^* is also reciprocal of the volume of the unit cell of the direct lattice and the relations between the angles are

$$\alpha^* = \sin^{-1}\{V/(abc \ \sin\beta \sin\gamma)\},$$
$$\beta^* = \sin^{-1}\{V/(abc \ \sin\alpha \sin\gamma)\},$$

and

$$\gamma^* = \sin^{-1}\{V/(abc \ \sin\alpha \sin\beta)\}.$$

6.6 The Interplanar Distance (d-Spacing) of Different Crystal Systems

The interplanar distance d_{hkl} in the direct lattice is related to the corresponding vector $\mathbf{H}_{hkl}$ in reciprocal space by the relation

$$d_{hkl} = 1/|\mathbf{H}_{hkl}|,$$

where $\mathbf{H}_{hkl} = h\mathbf{a}^* + k\mathbf{b}^* + l\mathbf{c}^*$. Therefore,

$$1/d_{hkl}^2 = (h\mathbf{a}^* + k\mathbf{b}^* + l\mathbf{c}^*) \cdot (h\mathbf{a}^* + k\mathbf{b}^* + l\mathbf{c}^*)$$
$$= h^2\mathbf{a}^* \cdot \mathbf{a}^* + k^2\mathbf{b}^* \cdot \mathbf{b}^* + l^2\mathbf{c}^* \cdot \mathbf{c}^* + 2hk\mathbf{a}^* \cdot \mathbf{b}^* + 2hl\mathbf{a}^* \cdot \mathbf{c}^* + 2kl\mathbf{b}^* \cdot \mathbf{c}^*,$$

where $\mathbf{a}^* = (\mathbf{b} \times \mathbf{c})/V$, $\mathbf{b}^* = (\mathbf{c} \times \mathbf{a})/V$, and $\mathbf{c}^* = (\mathbf{a} \times \mathbf{b})/V$. Putting these values of $\mathbf{a}^*$, $\mathbf{b}^*$, and $\mathbf{c}^*$ in the above expression for $1/d_{hkl}^2$, we get

$$1/d_{hkl}^2 = 1/V^2 \left\{ h^2|\mathbf{b} \times \mathbf{c}|^2 + k^2|\mathbf{c} \times \mathbf{a}|^2 + l^2|\mathbf{a} \times \mathbf{b}|^2 + 2hk(\mathbf{b} \times \mathbf{c}) \cdot (\mathbf{c} \times \mathbf{a}) \right.$$
$$\left. + 2hl(\mathbf{b} \times \mathbf{c}) \cdot (\mathbf{a} \times \mathbf{b}) + 2kl(\mathbf{c} \times \mathbf{a}) \cdot (\mathbf{a} \times \mathbf{b}) \right\}.$$

Now,

$$|\mathbf{b} \times \mathbf{c}|^2 = b^2c^2 \sin^2 \alpha, \quad |\mathbf{c} \times \mathbf{a}|^2 = c^2a^2 \sin^2 \beta, \quad \text{and} \quad |\mathbf{a} \times \mathbf{b}|^2 = a^2b^2 \sin^2 \gamma,$$

and further from vector product,

$$(\mathbf{b} \times \mathbf{c}) \cdot (\mathbf{c} \times \mathbf{a}) = (\mathbf{b} \cdot \mathbf{c})(\mathbf{c} \cdot \mathbf{a}) - (\mathbf{b} \cdot \mathbf{a})(\mathbf{c} \cdot \mathbf{c}) = abc^2(\cos \alpha \cdot \cos \beta - \cos \gamma),$$
$$(\mathbf{a} \times \mathbf{b}) \cdot (\mathbf{b} \times \mathbf{c}) = (\mathbf{a} \cdot \mathbf{b})(\mathbf{b} \cdot \mathbf{c}) - (\mathbf{a} \cdot \mathbf{c})(\mathbf{b} \cdot \mathbf{b}) = ab^2c(\cos \gamma \cdot \cos \alpha - \cos \beta),$$
$$(\mathbf{c} \times \mathbf{a}) \cdot (\mathbf{a} \times \mathbf{b}) = (\mathbf{c} \cdot \mathbf{a})(\mathbf{a} \cdot \mathbf{b}) - (\mathbf{c} \cdot \mathbf{b})(\mathbf{a} \cdot \mathbf{a}) = a^2bc(\cos \beta \cdot \cos \gamma - \cos \alpha).$$

Using these relations and factoring out $a^2b^2c^2$, we get

$$\frac{1}{d_{hkl}^2} = \frac{a^2b^2c^2}{V^2} \left\{ \frac{h^2 \sin^2 \alpha}{a^2} + \frac{k^2 \sin^2 \beta}{b^2} + \frac{l^2 \sin^2 \gamma}{c^2} + \frac{2hk}{ab}(\cos \alpha \cdot \cos \beta - \cos \gamma) \right.$$
$$\left. + \frac{2hl}{ac}(\cos \gamma \cdot \cos \alpha - \cos \beta) + \frac{2kl}{bc}(\cos \beta \cdot \cos \gamma - \cos \alpha) \right\}.$$

Now, V has been derived as

$$V = abc(1 - \cos^2 \alpha - \cos^2 \gamma - \cos^2 \beta + 2 \cos \alpha \cos \beta \cos \gamma)^{1/2}$$

and replacing this in the above expression, we get the d-spacing formulae as

$$\frac{1}{d_{hkl}^2} = \frac{1}{1 - \cos^2 \alpha - \cos^2 \gamma - \cos^2 \beta + 2 \cos \alpha \cdot \cos \beta \cdot \cos \gamma}$$
$$\times \left\{ \frac{h^2 \sin^2 \alpha}{a^2} + \frac{k^2 \sin^2 \beta}{b^2} + \frac{l^2 \sin^2 \gamma}{c^2} + \frac{2hk}{ab}(\cos \alpha \cdot \cos \beta - \cos \gamma) \right.$$
$$\left. + \frac{2hl}{ac}(\cos \gamma \cdot \cos \alpha - \cos \beta) + \frac{2kl}{bc}(\cos \beta \cdot \cos \gamma - \cos \alpha) \right\}.$$

This is the general expression for the d-value and is valid for *triclinic* system for which $\mathbf{a} \neq \mathbf{b} \neq \mathbf{c}$ and $\alpha \neq \beta \neq \gamma$. The expression becomes much simplified in the other systems having increasing symmetry like:

Monoclinic: $\mathbf{a} \neq \mathbf{b} \neq \mathbf{c}, \ \alpha = \gamma = 90° \neq \beta$

$$\frac{1}{d_{hkl}^2} = (1/\sin^2 \beta) \left(\frac{h^2}{a^2} + \frac{k^2 \sin^2 \beta}{b^2} + \frac{l^2}{c^2} - \frac{2hl \ \cos \beta}{ac} \right).$$

Orthorhombic: $\mathbf{a} \neq \mathbf{b} \neq \mathbf{c}, \ \alpha = \beta = \gamma = 90°$

$$\frac{1}{d_{hkl}^2} = \frac{h^2}{a^2} + \frac{k^2}{b^2} + \frac{l^2}{c^2}.$$

Rhombohedral: $\mathbf{a} = \mathbf{b} = \mathbf{c}, \ \alpha = \beta = \gamma \neq 90°$

$$\frac{1}{d_{hkl}^2} = \frac{(h^2 + k^2 + l^2) \sin^2 \alpha + 2(hk + kl + lh)(\cos^2 \alpha - \cos \alpha)}{a^2(1 + 2\cos^3 \alpha - 3\cos^2 \alpha)}.$$

Hexagonal: $\mathbf{a} = \mathbf{b} \neq \mathbf{c}, \ \alpha = \beta = 90°, \gamma = 120°$

$$\frac{1}{d_{hkl}^2} = \frac{4}{3}\left(\frac{h^2 + hk + k^2}{a^2}\right) + \frac{l^2}{c^2}.$$

Tetragonal: $\mathbf{a} = \mathbf{b} \neq \mathbf{c}$, $\alpha = \beta = \gamma = 90°$

$$\frac{1}{d_{hkl}^2} = \frac{h^2 + k^2}{a^2} + \frac{l^2}{c^2}.$$

Cubic: $\mathbf{a} = \mathbf{b} = \mathbf{c}$, $\alpha = \beta = \gamma = 90°$

$$\frac{1}{d_{hkl}^2} = \frac{h^2 + k^2 + l^2}{a^2}.$$

6.7 Weighted Reciprocal Lattice

It has been explained before how the concept of reciprocal lattice is important particularly in describing the diffraction pattern of a crystal. This purpose will be more fruitfully served if each lattice point in the reciprocal space can also

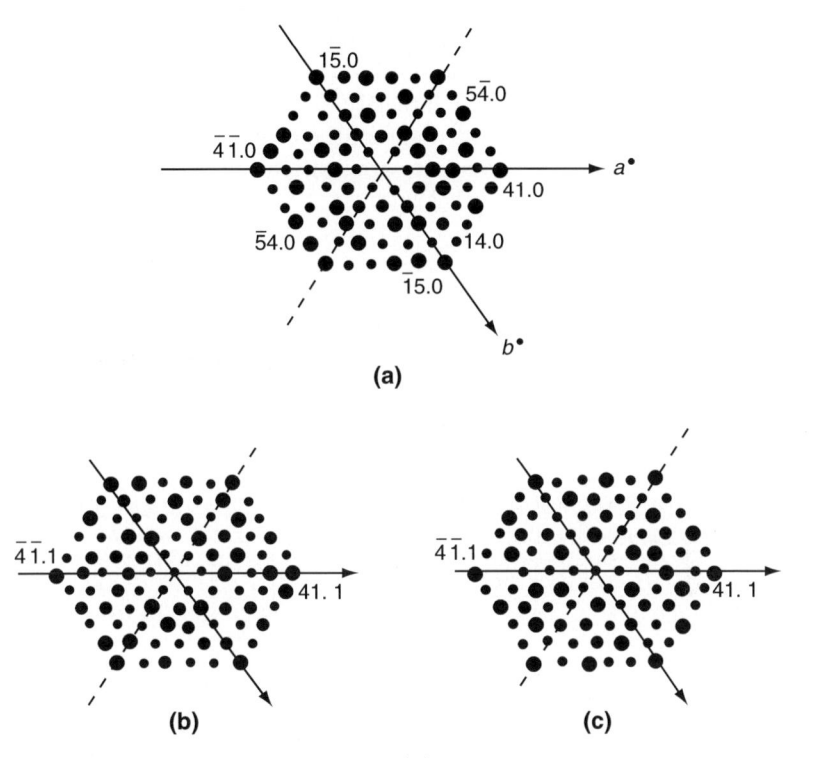

Fig. 6.10. Weighted reciprocal lattice for (**a**) zero layer hexagonal or rhombohedral, (**b**) first layer rhombohedral, and (**c**) first layer hexagonal (By courtesy of L.S. Dent Glasser)

represent the intensity of diffraction from the corresponding plane in the direct lattice. This is achieved by assigning to each point a weight proportional to the intensity of the corresponding reflection. The resulting *weighted reciprocal lattice* is usually drawn by making each point with a dot whose size is roughly proportional to the observed intensity. Weighting the points in this way brings out the relationship between the symmetry of the crystal and the symmetry of its diffraction pattern. Figure 6.10 shows the weighted reciprocal lattice of a hexagonal crystal. However, the discussion of the utility of this weighted reciprocal lattice will be made in a later chapter.

6.8 Summary

1. The basics of X-ray production, its monochromatization and diffraction from crystal, and the diffraction laws are discussed.
2. On the basis of the crystal structure analysis, the concept of reciprocal lattice has been introduced and the importance of its concept has been emphasized.
3. The expressions of the interplanar distances and also the mathematical relations between the hypothetical reciprocal lattice and the real direct lattice parameters have been derived and shown.
4. There exists an important relationship between the position of an atom or molecule in the lattice and the intensity of the diffracted X-radiation, and to incorporate the both in one diagram the concept of weighted reciprocal lattice is introduced here and its applications in structure analysis are discussed in appropriate chapter later.

References

1. A.H. Compton, S.K. Allison. *X-Ray in Theory and Experiment* (D. Van Norstand, New York, 1935)
2. W.T. Sproull, *X-Ray in Practice* (McGraw Hill, New York, 1942)
3. B.D. Cullity, *Elements of X-Ray Diffraction* (Addison-Wesley, Reading, MA, 1978)
4. S.K. Chatterjee, *X-Ray Diffraction: Its Theory and Applications* (Prentice Hall, New Delhi, 1999)

7

Experimental Methods for Structure Analysis: X-Ray Diffraction Techniques

A crystal is made up by arranging the corresponding unit cells in three directions. If there is an order in the repetition of this unit cell arrangement in space, then only its crystalline nature is manifested. This order is of two types (a) local order and (b) long-range order. When both of these two types of ordering are maintained in space then the crystal is called *single crystal*, and when the local order is maintained but the long-range order is violated often then the crystalline material is called *polycrystal*. The region within which these two types of ordering are maintained but the long-range ordering is violated just on its boundary is single crystal region. The size of these regions behaving like single crystals depends on the resolution power of probe which is used to "see" them. The X-ray diffraction patterns from these two types of crystals are obviously different and so also the experimental techniques to get them. In this chapter, the different experimental techniques and the method of their interpretations generally used for the structural studies are discussed for both of these two types of crystalline state of matter, i.e., single and polycrystalline states [1–3].

7.1 Experimental Techniques for Single Crystal

There are two types of experimental techniques for single crystal studies. This classification is made on the basis of the wavelength λ and Bragg's angle θ as:

Type (A). Bragg's angle θ is known but the wavelength λ which satisfies the Bragg condition is unknown.

Type (B). Bragg's angle θ as well as the wavelength λ which satisfy the Bragg condition are known.

7.1.1 Laue Camera and Laue Pattern

Laue camera is the simplest of all the devices for the structure determination of crystals as it consists of a collimator to narrow down the general radiation

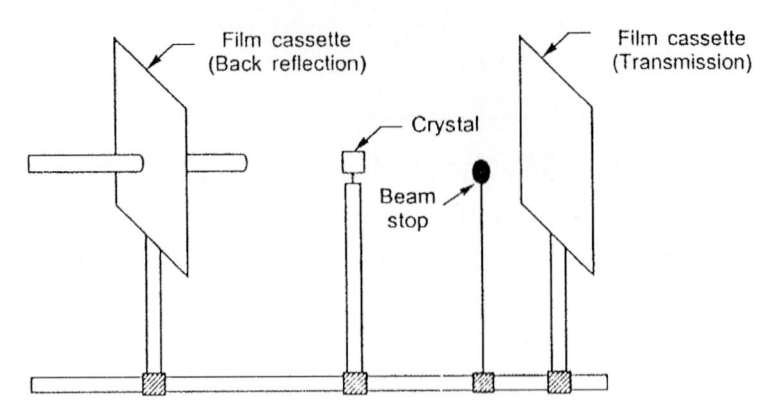

Fig. 7.1. Laue pin-hole camera. Single crystal is stationary and two flat fixed films for recording transmission (possible if the crystal is transparent to X-ray) and back reflection (for large crystals)

of X-ray, a two circle goniometer to mount the single crystal and align its desired axis perpendicular to the incident beam and a beam outlet at the back of which there is a fluorescent screen to see the transmitted beam followed by a lead glass absorber. There are two film cassettes that are flat and one of them is mounted through the inlet collimator and the other in the transmitted side. In addition to these, there is one lead stop to stop the transmitted direct beam from reaching the film. All these, i.e., the film cassettes, beam stop, and the goniometer are mounted on a bridge so that their distances are adjustable. Figure 7.1 shows the schematic diagram of a Laue camera.

A Laue pattern which is shown in Fig. 7.2 is symmetrical arrangement of diffraction spots on the flat film either in the transmission mode or in back reflection mode. The spots when connected generate either an ellipse or a hyperbola. These spots originate due to Bragg's reflection from sets of intersecting planes which are parallel to an axis known as *zone axis* (Fig. 7.3). Planes having common zone axis generate symmetrical spots, and in most of the cases they show a center of symmetry.

Let us introduce any vector $\mathbf{A}(uvw)$ as $\mathbf{A}(uvw) = u\mathbf{a}+v\mathbf{b}+w\mathbf{c}$, where uvw are integers. All planes containing the direction $\mathbf{A}(uvw)$ are said to belong to the same zone $u\ v\ w$. Now, $\mathbf{H}_{hkl}$ which is defined as $\mathbf{H}_{hkl} = h\mathbf{a}^* + k\mathbf{b}^* + l\mathbf{c}^*$ is a vector normal to the plane hkl.

Therefore,

$$\mathbf{A}(uvw) \cdot \mathbf{H}_{hkl} = (u\mathbf{a} + v\mathbf{b} + w\mathbf{c}) \cdot (h\mathbf{a}^* + k\mathbf{b}^* + l\mathbf{c}^*)$$
$$= hu + kv + lw = 0$$

as the vector $\mathbf{A}(uvw)$ lies on the plane and $\mathbf{H}_{hkl}$ perpendicular to it. So, the equation $hu + kv + lw = 0$ represents the planes hkl which belong to the zone

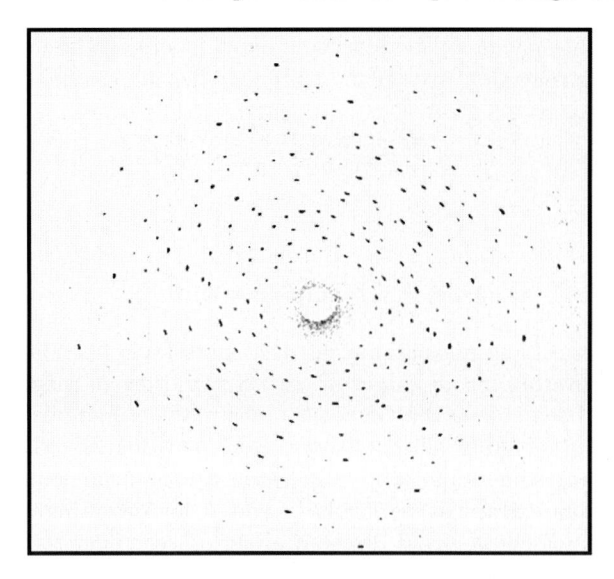

Fig. 7.2. Transmission Laue photograph, showing diffraction spots arranged symmetrically (By courtesy of L.S. Dent Glasser)

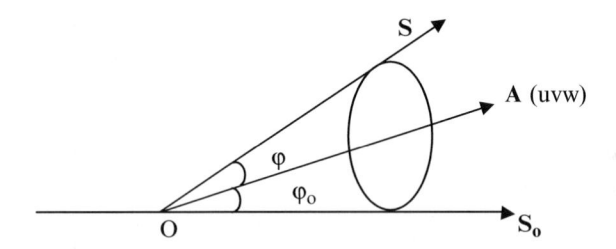

Fig. 7.3. Zone axis $\mathbf{A}(uvw)$, $\mathbf{S}$ and $\mathbf{S}_0$ are the unit vectors defining, respectively, the diffracted and incident beam directions

$u\ v\ w$. Now, as $\mathbf{S} = \mathbf{S}_0 + \lambda\mathbf{H}_{hkl}$ (from Ewald's sphere), multiplying both sides by $\mathbf{A}(uvw)$ we get

$$\mathbf{S} \cdot \mathbf{A}(uvw) = \mathbf{S}_0 \cdot \mathbf{A}(uvw) + \lambda\mathbf{H}_{hkl} \cdot \mathbf{A}(uvw), \quad \text{and as} \quad \mathbf{H}_{hkl} \cdot \mathbf{A}(uvw) = 0,$$

we get

$$|\mathbf{A}(uvw)| \cos\varphi = |\mathbf{A}(uvw)| \cos\varphi_0 \quad \text{and so,} \quad \varphi = \varphi_0.$$

Note. Therefore, diffracted beams from the sets of hkl planes form a cone and the zone axis is the axis of that diffraction cone and all such sets of planes belong to the same zone $u\ v\ w$ defined by the zone axis $\mathbf{A}(uvw)$.

There are some difficulties in interpreting the Laue pattern in terms of identifying the reflected spots on the film. We recall the Bragg's law as

$$2d_{hkl} \sin\theta = n\lambda \quad \text{and} \quad \sin\theta/n = \lambda/2d_{hkl},$$

where d_{hkl} is the spacing of the reflecting plane having Miller indices hkl, now considering the reflection from plane having Miller indices nh, nk, nl we get

$$2d_{nh,nk,nl} \sin \theta' = \lambda$$

and

$$\sin \theta' = \lambda/2d_{nh,nk,nl} \quad \text{but as} \quad d_{nh,nk,nl} = d_{hkl}/n,$$
$$= n\lambda/2d_{hkl} = n \cdot \sin \theta/n = \sin \theta.$$

Therefore, in the Laue photograph, all the spots for reflection from nh, nk, nl planes will overlap for all values of n with nth order of reflection from hkl planes. It has been shown that in most realistic experimental arrangements, 95% of the spots should be either without overlap or singly overlapped and the reason for having some Laue spots high intensity than the neighboring ones.

The Laue photograph is particularly suited for detecting the symmetry present if the incident X-ray is directed along or very near to the symmetry axis. The diffraction spots will show that rotational symmetry in their position on the film as the X-ray beam direction represents in the original crystal. If the beam direction represents accurately say the triad axis of symmetry present, then the diffraction spots will form equilateral triangle otherwise even if the beam direction is slightly offset from the triad axis of symmetry, the trigonal symmetry will still be evident from the triangles of spots which will not be equilateral [2, 3].

7.1.2 Rotation/Oscillation Camera and the Applications

In this arrangement, the crystal is a single crystal as in Laue arrangement but it is either rotated or oscillated through certain angular range, the X-ray is monochromatic and the film instead of being flat is cylindrical. Therefore, the camera arrangement is also called *cylindrical camera*. Figure 7.4 shows the arrangement of a cylindrical camera. Here, the film though cylindrical is kept stationary as in Laue.

As X-ray beam is monochromatic say K_α so the wavelength is known. The diffraction cones originating from the crystal will form layers as shown in the following rotation/oscillation X-ray photograph (Fig. 7.5).

Let $\mathbf{S}$ and $\mathbf{S}_0$ be the diffracted and incident beam directions and $\mathbf{a}_3$ be the crystal axis about which the crystal is mounted in Fig. 7.4. Then from Laue condition,

$$(\mathbf{S} - \mathbf{S}_0) \cdot \mathbf{a}_3 = l\lambda \quad \text{but as} \quad \mathbf{S}_0 \cdot \mathbf{a}_3 = 0,$$
$$\mathbf{S} \cdot \mathbf{a}_3 = l\lambda \quad \text{or,} \quad \mathbf{a}_3 \sin \beta = l\lambda \quad \text{or,} \quad \sin \beta = l\lambda/\mathbf{a}_3,$$

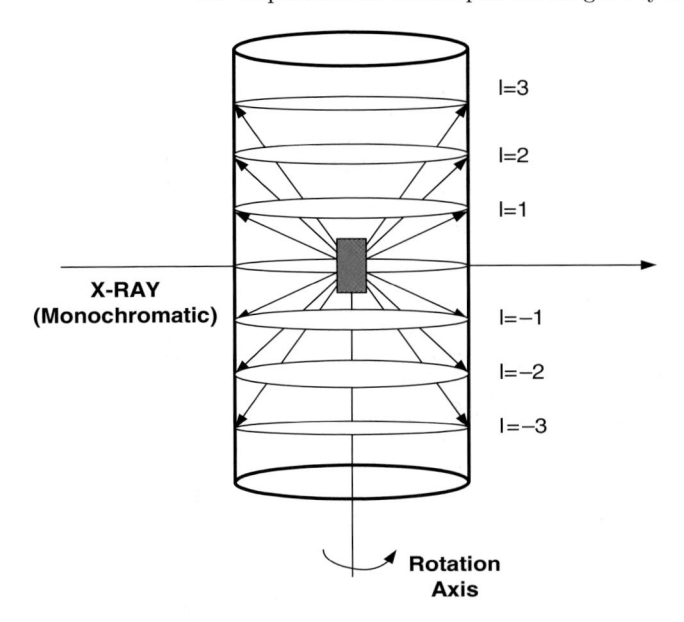

Fig. 7.4. The rotation–oscillation (cylindrical) camera. Monochromatic K_α is used. The crystal shown at the center is rotated or oscillated. The layer lines formed by the intersection of diffraction cones and the cylindrical film are shown for different l values

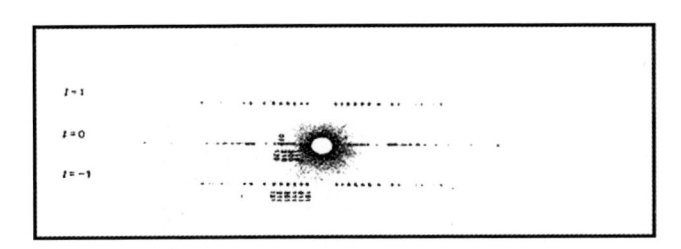

Fig. 7.5. The oscillation photograph of a single crystal mounted along its c-axis (By courtesy of L.S. Dent Glasser)

where β is the latitude angle and reflections having same value of l will have the same latitude angle and form a diffraction cone. When cylindrical film is used, the cones form layers on the film, for each value of, i.e., 0, 1, 2, -1, -2, etc. (see Fig. 7.6).

Now, measuring the distance y_n of the nth layer from the equatorial layer ($l = 0$), we get

$$\tan \beta_n = y_n/R,$$

where R is the camera radius.

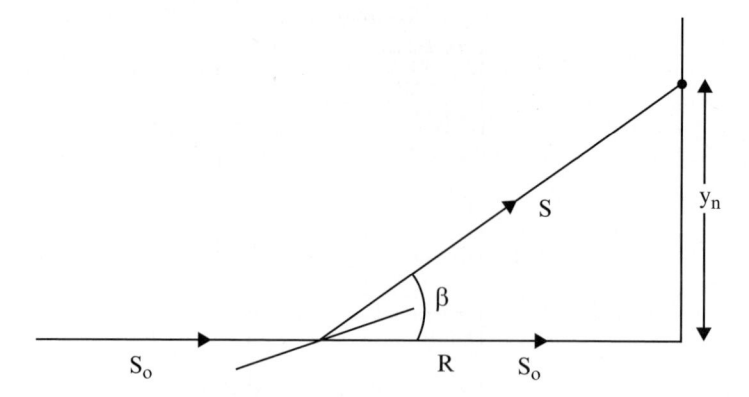

Fig. 7.6. Diffraction of X-ray in rotation/oscillation camera

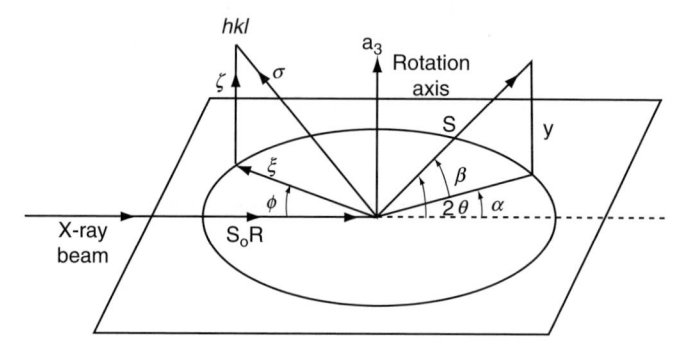

Fig. 7.7. The relation between cylindrical coordinates and indices of reciprocal lattice. Points of rotation photograph (indexing of rotation photographs)

The axial lengths of a crystal $\mathbf{a}_1$, $\mathbf{a}_2$, or $\mathbf{a}_3$ can be determined by mounting the crystal in the corresponding axis and measuring it from rotation photographs. This measurement does not require any knowledge of the crystal system or hkl indices of the reflections. Now, to find the correlation between reciprocal lattice points and diffraction spots on the film, let us consider Fig. 7.7 [2].

Let ζ, ε, and φ be the cylindrical coordinates of the reciprocal lattice point hkl with respect to the rotation axis and direct beam. Now, if σ represents the position vector of the reciprocal lattice point, then $\sigma = \zeta + \varepsilon$.

We know from Bragg's law

$$\mathbf{S} - \mathbf{S}_0 = \lambda \mathbf{H} = \lambda \sigma.$$

Now taking the vertical components of the vectors $|\mathbf{S}_1| - |\mathbf{S}_{01}| = \lambda|\mathbf{H}_1| = \lambda\zeta$, but $\mathbf{S}_{01}$ is zero as $\mathbf{S}_0$ lies on the plane and perpendicular component $\mathbf{S}_1$ of the unit vector $\mathbf{S}$ is $\sin\beta$.

Therefore, $\sin\beta = \lambda|\mathbf{H}_1| = \lambda\zeta$ and the vertical coordinate y of the spot on the film is given by

$$y = R \tan \beta = \frac{R}{\sqrt{(1/\lambda\zeta)^2 - 1}}. \tag{7.1}$$

Now, the angle α in the Fig. 7.6 is given by $\alpha = x/R$, where x is the horizontal coordinate of the film, then from spherical triangle $\cos 2\theta = \cos \alpha \cdot \cos \beta$, we get

$$\cos\left(\frac{x}{R}\right) = \frac{\cos 2\theta}{\cos \beta} = \frac{1 - 2 \sin^2 \beta}{\sqrt{1 - \sin^2 \beta}} = \frac{1 - 1/2(\lambda/d_{hkl})^2}{\sqrt{1 - (\lambda\zeta)^2}} = \frac{1 - 1/2(\lambda H_{hkl})^2}{\sqrt{1 - (\lambda\zeta)^2}}$$

$$= \frac{1 - 1/2(\lambda\sigma)^2}{\sqrt{1 - (\lambda\zeta)^2}} = \frac{1 - 1/2[(\lambda\zeta)^2 + (\lambda\xi)^2]}{\sqrt{1 - (\lambda\zeta)^2}}. \tag{7.2}$$

Now, measuring the vertical coordinate y from the film of the spot, one can get from (7.1) the ζ coordinate (provided the camera radius R and wavelength λ are known) and knowing ζ, the value of ξ can be found out after measuring the horizontal coordinate x and using the (7.2). These calculations are, however, avoided by using the Barnal chart relating x, y with ζ, ξ. The Barnal chart in the form of a plot of coordinates x, y with ζ, ξ for different standard camera radius is commercially available on transparent paper, and the coordinates can be easily read out by superimposing the rotation photographs over the such transparent charts. However, if the needle-shaped crystal is mounted on its c-axis, then from Laue condition

$$c \cos \theta = 1\lambda \quad \text{and} \quad \cos \theta = \zeta/\lambda^{-1}$$

then

$$c = 1\lambda/\zeta\lambda = 1/\zeta.$$

Here, 1λ is the radius of the sphere of reflection, and thus mounting the crystal in two other directions as rotation axis, the other two unit translational vectors **a** and **b** can be determined [2, 3].

Now, as the values of ζ and ξ, respectively, give the vertical distance from the zero layer of reciprocal net and the horizontal distance from the rotation axis, assigning the h, k indices to the spots within a layer is usually not a straightforward process. The problem is that all the information from two-dimensional reciprocal lattice layer is compressed into the one-dimensional layer – line. As an exercise, it may be tried by first drawing the reciprocal lattice net and then drawing arcs of circles from the origin of the net having radii equal to the ξ values. The point of intersections of such arcs with the reciprocal lattice points on the net gives the probable values of h and k for different layers. However, it becomes increasingly difficult to assign indices to the spots at higher value ξ. This problem can be decreased by taking an oscillation photograph when crystal is oscillated through certain angular range instead of taking a complete rotation photograph. The amount of information recorded is thus reduced and thus interpretation is made correspondingly easier. For crystals with low symmetry and/or large unit cells, it is usually not worthwhile to sort this out but try to spread this information out into two dimension using a moving film camera like Weissenberg camera. However, the main advantages or importance of an oscillation method are that first it gives the measurement of crystal axes like **a, b**, and **c** as described above and secondly it is essential for the alignment of the crystal axis along the rotational axis of the camera. If the crystal is needle shaped, it becomes difficult to align the symmetry axes other than the needle axis along the rotation axis. In case the rotation axis differs, then the rotation/oscillation layer line will appear wavy instead of being straight lines. There exist standard methods for aligning the crystal by measuring the distance between the existing wavy layer line and the expected layer line. The details of this process of alignment, however, are avoided in this book and can be found in any book dealing the details of photographic techniques.

7.1.3 Weissenberg Camera and Moving Film Technique

It should be mentioned here that the reciprocal lattice which is three dimension cannot be recorded without any ambiguity on a two-dimensional film and so, the rotational method discussed before is not sufficient to record the diffraction spots with their three hkl indices known. Therefore, the rotational method cannot give any satisfactory information about the Space Group from the rule of systematic absences. Moreover, there is every chance of coincidence of the diffraction from one hkl and its $\overline{hkl}$ counterpart and so, they hardly can be read separately. Therefore, it becomes necessary to move the film along with the rotation of the crystal to record one reciprocal lattice net lying on a plane.

There are two types of moving film cameras: Precision and Weissenberg cameras. Of these two, Weissenberg camera is closer to rotation/oscillation camera. But former of them records the undistorted reciprocal lattice while the latter records the distorted one. The Weissenberg camera which is discussed here only records relatively smaller part of reciprocal lattice as the crystal is not rotated through entire range of $360°$ but instead oscillated through an angular range say $120°$, and the film whose linear movement is perfectly coupled with the oscillation of the crystal is also translated back and forth through certain length of distance. This is to avoid the major drawback that is found in simple rotation/oscillation camera, i.e., recording two-dimensional reciprocal lattice net of certain value of l into one-dimensional layer, which results almost impossibility of assigning the indices to the reflected spots without any ambiguity. To preserve more certainty in the process of assigning indices, a particular layer line is selected at a time in Weissenberg by using a layer line screen of cylindrical shape and having a slot perpendicular to its axis. By adjusting the slot opening in the position of the desired layer and rotating the crystal through certain precalculated angle, the incident beam coincides with the diffraction cone of the desired layer and the diffracted cone can reach the translating film through the aligned slot on the layer screen. Such arrangement in Weissenberg camera is called *equi-inclination setting*. Figure 7.8 shows an arrangement of the Weissenberg camera [3].

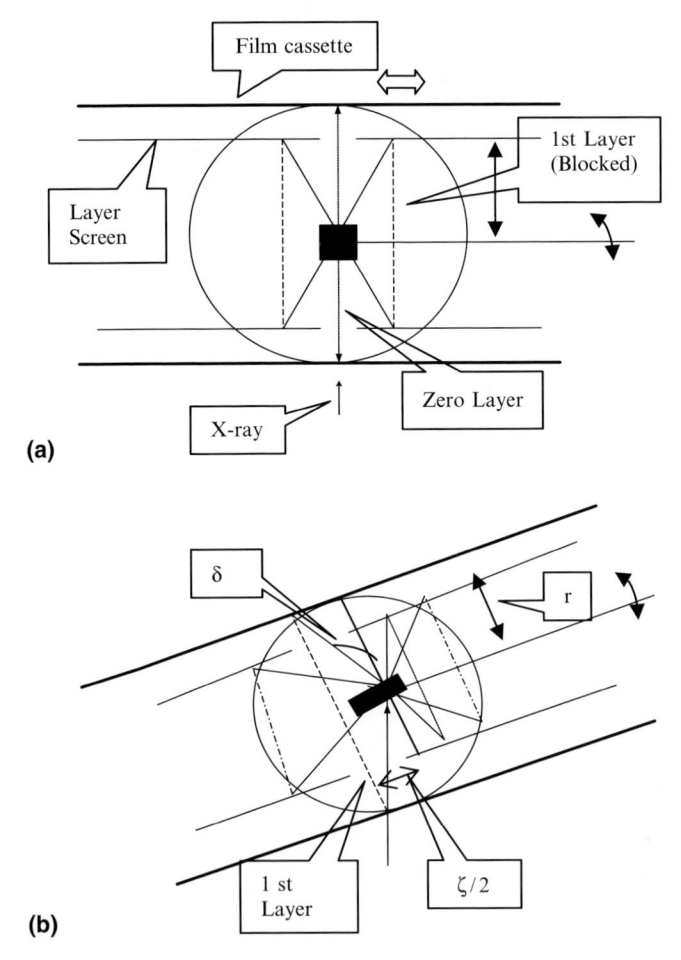

Fig. 7.8. Equi-inclination Weissenberg camera. (**a**) is the arrangement for recording the *zero layer* and (**b**) is the arrangement for recording the *first layer photographs*. *Note*: in (**b**), the crystal and the film cassette with the layer screen are rotated through an angle δ and the layer line selector screen is shifted through a distance of $\zeta/2 = r \tan \delta$ for recording the first layer diffraction cone. The incident X-ray beam coincides with the diffraction cone in this equi-inclination setup

The result of the oscillation of crystal and translational motion of the film which is synchronized with the oscillatory motion of the crystal is highly distorted picture of the reciprocal lattice. While the distance of any spot from the equator depends on ξ, its horizontal distance along the film represents the angular position of the crystal as the corresponding reciprocal lattice point passes through the surface of the sphere of reflection (Fig. 7.9). Figure 7.10 indicates how this distortion can be visualized and why this equi-inclination

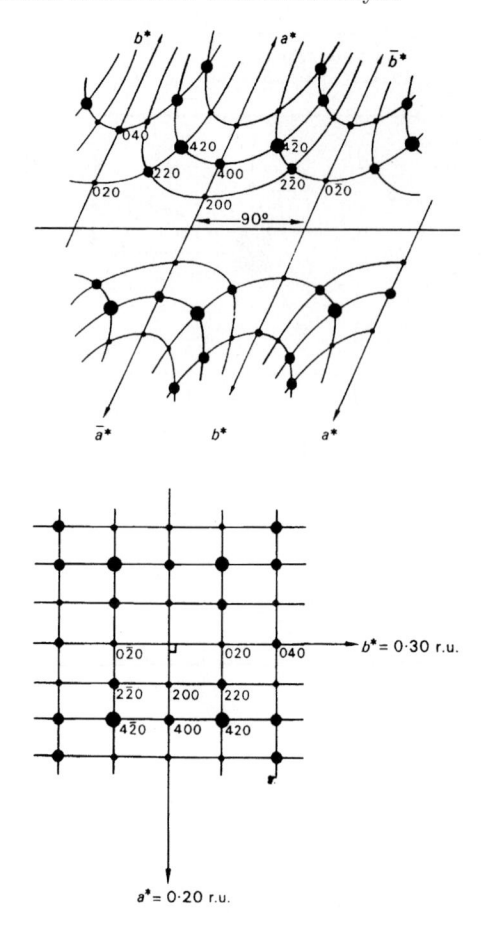

Fig. 7.9. Replica of zero layer Weissenberg photograph and the undistorted weighted reciprocal lattice (By the courtesy of L.S. Dent Glasser)

Weissenberg is said to record the distorted reciprocal lattice separately for each layer of reciprocal lattice net. Imagine the weighted reciprocal lattice layer drawn on a stretchable plane, with a rod inserted along each axis to keep it straight. These axial rods are then pulled apart at the origin until they are parallel. The axes are then inclined to the equator because to bring successive axial points into sphere of reflection the reciprocal lattice has to rotate. The quantitative interpretation is done with the help of Weissenberg chart, which enables coordinates in reciprocal space to be read directly from the photograph.

Now, knowing such hkl values of the spots of the X-ray diffraction photographs which are undistorted record of the reciprocal lattice, the Space Group and also the crystal structure can be determined. It may be once again stated here that by Space Group, we should mean the spatial symmetries which

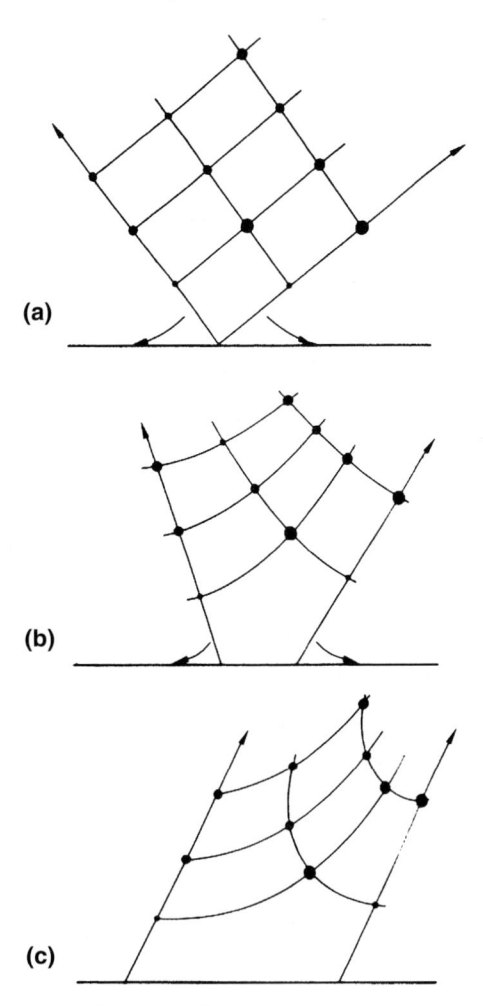

(a)

(b)

(c)

Fig. 7.10. (a) The weighted reciprocal lattice drawn say on a stretchable surface. (b) The axes are drawn from the origin away from each other with an intention to make them parallel to each other. (c) The axes are made parallel. The diffraction spots then resemble the Weissenberg photograph (replica) of Fig. 7.9, showing streamers and festoons (By courtesy of L.S. Dent Glasser)

are present in three directions of the unit cell amongst the different atoms present in the unit cell. By crystal structure, we should mean knowing fully the positions of all the atoms present in the unit cell. Therefore, determination of the Space Group is usually the step which precedes the process for structure determination. A discussion rather in brief is given in the following chapter which is followed for Space Group and structure determination.

7.1.4 de Jong–Boumann and Precession Camera

It has been said that the Precision camera records the undistorted reciprocal lattice. There are two types of Precision cameras. The one is due to de Jong and Boumann and the other is due to Buerger. The Precession camera developed by Buerger is quite different and difficult to visualize and so, the Boumann method is first described here.

The layer screen used in this method is annular opening in which the central opaque part is supported to the outer part, leaving the desired annular opening by cellophane paper which is transparent to X-ray (Fig. 7.11). In de Jong and Boumann method, like Weissenberg, the diffraction data are spread out over the flat film by rotating it (Figs. 7.12 and 7.13).

If μ and ν are the angles, respectively, between the X-ray incident direction and crystal axis and semiapex angle of the diffraction cone as shown in the following figure,

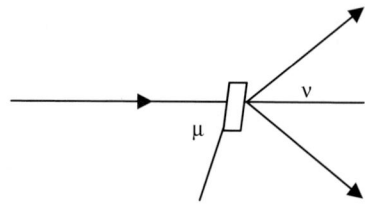

Then, the distance through which the film is to be moved back and the formulae relating to the recording of upper layer are given as: For zero layer arrangement $\mu = \nu = 45°$, layer screen subtends $90°$ cone at the crystal, and this is to be kept constant. The upper layer is to be recorded by changing μ. Now for nth layer,

$$\sin \mu_n = \cos \nu - \frac{n\lambda}{t} = \frac{1}{\sqrt{2}} - \frac{n\lambda}{t}, \quad \text{since } \mu \text{ is fixed.}$$

The film is to be moved away from the crystal by

$$\frac{F \cdot n\lambda}{t},$$

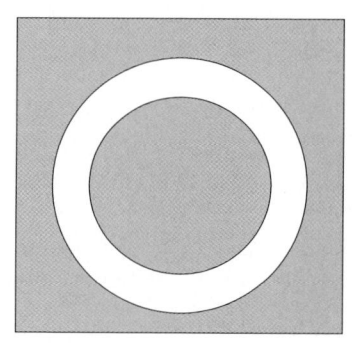

Fig. 7.11. The annular opening in layer separator slit

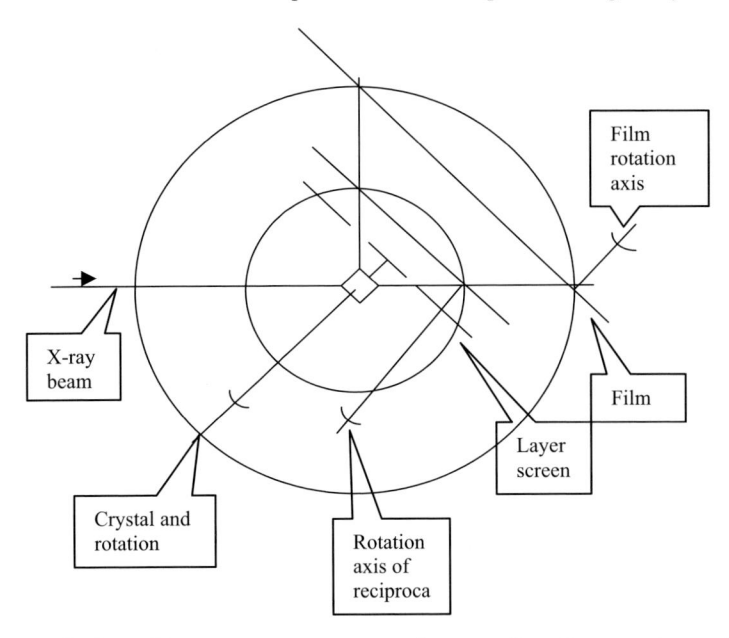

Fig. 7.12. de Jong–Boumann camera setup for recording the zero layer. (**a**) The crystal is mounted with its principal axis perpendicular to the film and rotated along its axis. The diffracted beam makes 90° with the incident beam. (**b**) The zero layer rotates about the parallel axis. (**c**) The flat film parallel to the set of reciprocal lattice layers is also rotated about the axis. (**d**) The rotation of film and reciprocal lattice is synchronized, so that the diffracted spot reaches the same point on the film

where F is the crystal to film distance and t is the unit cell translation along the relevant principal axis of the crystal.

Precession camera. It has been stated that the visualization of Precession camera is rather difficult compared with de Jong–Boumann camera. Therefore, only zero layer Precision setup is described in brief. The detailed information is available in more specialized books given in the references for further reading [2].

While de Jong–Boumann method records reciprocal lattice layers perpendicular to the axis of the goniometer head, the precession method records layers that are parallel to it. Thus, the de Jong method for orthogonal crystal mounted about c can record $hk0$, $hk1$, etc., layers, and the precession method records both $0kl$, $1kl$, etc., and $h0l$, $h1l$, etc., layers.

The chief drawbacks to both of these two methods are that both of them record only a small part of the reciprocal lattice. This may not be of any difficulty if the crystal has short reciprocal dimension, but for crystals with larger reciprocal dimensions too little information may be recorded for the systematic absences to be fully investigated. One way to escape from this is to use X-ray

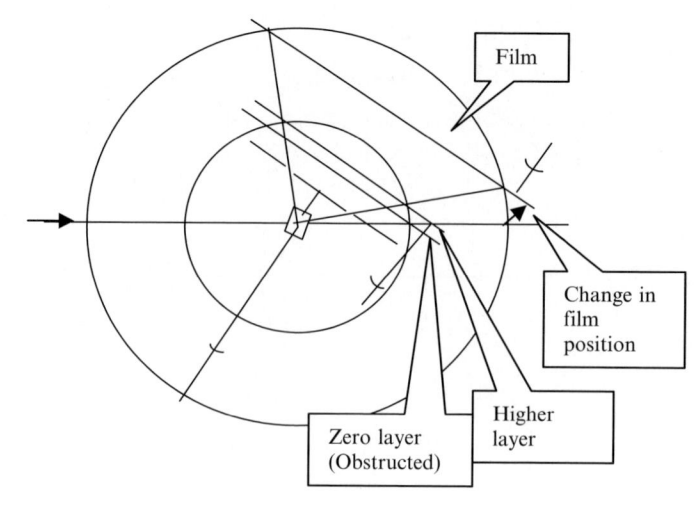

Fig. 7.13. de Jong–Boumann camera setup for recording the higher layer. (**a**) The crystal is rotated and so also its reciprocal lattice without altering the relative positions of the crystal, screen, and rotation axis of the film remains unchanged. (**b**) The film is moved back along its rotation axis by a distance proportional to the height of the nth layer above zero layer so as to record the higher layer. (**c**) The diffraction from the zero layer is obstructed by the layer screen. (**d**) The upper layer is brought into recording position by altering the angle of incidence of X-ray beam

of shorter wavelength, but this may also have certain disadvantages and in that condition one may have to depend upon other methods like Weissenberg which records distorted reciprocal lattice (Fig. 7.14).

7.2 Experimental Techniques for Polycrystals

A polycrystal is in one way an unsymmetrical state of crystalline matter compared with single crystals. The long-range order which is extended all through the bulk is maintained but the short-range order is often violated. The region or the plane where this long-range order is violated is known as the *grain boundary*. Within a grain, each crystal plane is oriented in a particular fashion whereas this orientation order is not obeyed at the grain boundary. In a perfect polycrystal, each crystal plane is randomly oriented or rather it is oriented in all possible fashion. The result is the total change of diffraction pattern, in place of getting diffraction spots we get a cone of diffraction for a particular wavelength. Figure 7.15 shows the diffraction cone resulting from a polycrystalline specimen.

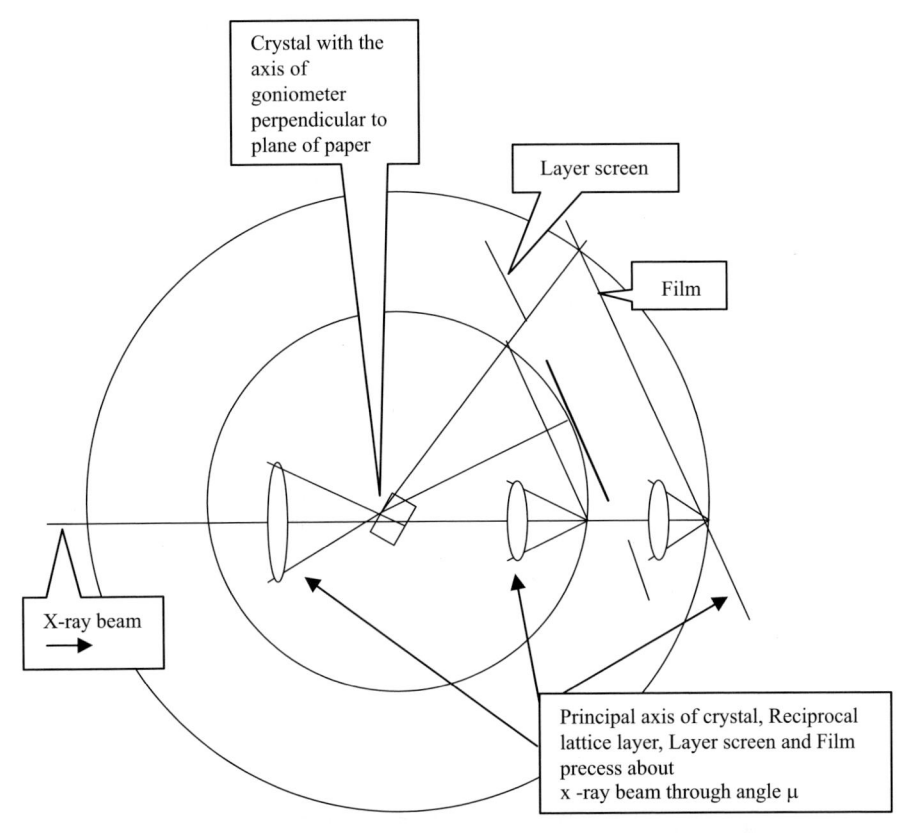

Fig. 7.14. Zero layer setup for Buerger precession camera

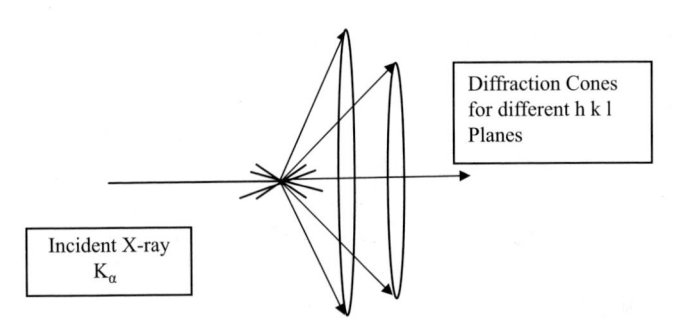

Fig. 7.15. Diffraction cones originating from different planes for the same X-ray wavelength. Different Bragg's angle for different planes present in different grains

7.3 Debye–Scherrer Cylindrical Powder Camera: The Plan View

$$2\theta_{rad} = \frac{2S}{2R}, \quad 2\theta^0 = 180/\pi \times 2\theta_{rad}$$

and from Bragg's law

$$2d_{hkl} \sin \theta = n\lambda.$$

So, measuring $2S$ from film which is loaded on the inside wall and knowing R the radius of the camera and the wavelength λ, we can calculate the inter planar spacing d_{hkl} for the hkl plane. The d_{hkl} for the cubic system is related to the lattice parameter a by

$$d_{hkl} = \frac{a_{hkl}}{\sqrt{h^2 + k^2 + l^2}}.$$

By using Debye–Scherrer technique for polycrystals, lattice parameter can be determined. The lines (arcs) can be indexed by adopting the following procedures (Fig. 7.16) [1].

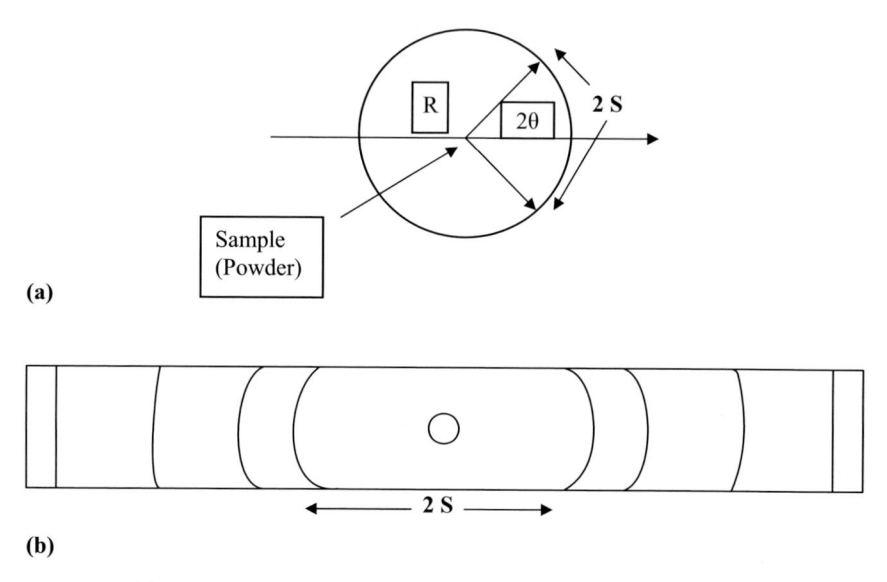

Fig. 7.16. (a) The Debye–Scherrer cylindrical camera and incident monochromatic K_α radiation. (b) The cylindrical film and Debye arcs from various planes

7.4 Indexing of the Debye–Scherrer Pattern

7.4.1 Cubic Systems

We know that from Bragg's law, $2d \sin \theta = n\lambda$. And, for cubic system $1/d^2 = (h^2 + k^2 + l^2)/a^2$ and from Bragg's law $1/d^2 = 4\sin^2 \theta/n^2\lambda^2$, for $n = 1$, we get $1/d^2 = 4\sin^2 \theta/\lambda^2$. Now, equating two $1/d^2$ values,

$$\frac{\sin^2 \theta}{h^2 + k^2 + l^2} = \frac{\sin^2 \theta}{s} = \frac{\lambda^2}{4a^2},$$

where $s = h^2 + k^2 + l^2$ which is a constant for any one particular pattern.

Now, the problem of indexing the pattern of a cubic system turns out to find the constant quotient when observed $\sin^2 \theta$ values are divided one after another by all combinations of $h^2 + k^2 + l^2$ like $1, 2, 3, 4, 5, 6, \ldots$ formed by the sum of three squared integers. Each of the four common cubic lattice types has characteristic sequence of diffraction lines, given by their sequential s values:

Simple cubic: $1, 2, 3, 4, 5, 6, 8, 9, \ldots$ (all reflections)
Body-centered cubic: $2, 4, 6, 8, 10, 12, \ldots (h + k + l = 2n)$
Face-centered cubic: $3, 4, 8, 11, 12, \ldots (h, \ k, \ l$ are either all even or all odd)
Diamond cubic: $3, 8, 11, 16, \ldots$

Now, each set should be tried in turn and if any set of integers is not found giving constant $\lambda^2/4a^2$ value, then it should be concluded that the sample does not belong to cubic system and other possibilities like that of tetragonal, hexagonal, etc., are to be explored.

7.4.2 Tetragonal System

$$\frac{1}{d^2} = \frac{h^2 + k^2}{a^2} + \frac{l^2}{c^2}$$

and from Bragg's law

$$\frac{1}{d^2} = \frac{4\sin^2 \theta}{\lambda^2}$$

and equating the RHS, we get

$$\sin^2 \theta = \frac{\lambda^2}{4a^2}(h^2 + k^2) + \frac{\lambda^2}{4c^2}l^2.$$

Now, for $l = 0$, i.e., for $hk0$ reflections $h^2 + k^2 = 1, 2, 4, 5, 8, \ldots$ in the equation $\lambda^2/4a^2 = [1/(h^2 + k^2)]\sin^2 \theta$, a search for finding constant values for $\lambda^2/4a^2$ satisfying integers and the corresponding reflections are to be sorted out and the rest reflections are then from hkl reflections for which $l \neq 0$. The above equation can then be written as

$$\sin^2 \theta - A(h^2 + k^2) = Cl^2,$$

where $A = \lambda^2/4a^2$ and $C = \lambda^2/4c^2$. Now, for various combination of h, k for LHS giving RHS values in the ratio of $1, 4, 9, 16, \ldots$ are sorted out and corresponding hkl are identified. c the second lattice parameter is then calculated from the ratio 1 and then confirmed from $4, 9, \ldots$

7.4.3 Hexagonal System

For this system,

$$\frac{1}{d^2} = \frac{4}{3}\left(\frac{h^2 + hk + k^2}{a^2}\right) + \frac{l^2}{c^2}$$

and from Bragg's law $1/d^2 = 4\sin^2 \theta/\lambda^2$ and equating both, we get

$$\sin^2 \theta = \frac{\lambda^2}{3a^2}(h^2 + hk + k^2) + \frac{\lambda^2}{4c^2}l^2, \quad \text{and putting } l = 0,$$

$$\sin^2 \theta = \frac{\lambda^2}{3a^2}(h^2 + hk + k^2) \quad \text{and} \quad \frac{\lambda^2}{3a^2} = \frac{\sin^2 \theta}{h^2 + hk + k^2} = \frac{\sin^2 \theta}{1, 3, 4, 7, 9, \ldots}.$$

Dividing each $\sin^2 \theta$ values sequentially by integers $1, 3, 4, 7, 9, \ldots$, the match for $\lambda^2/3a^2$ giving constant values is then indexed and the indexed $(hk0)$ lines are to be sorted out and the lattice parameter a is to be calculated. For rest of the lines $(l \neq 0)$, we get

$$\sin^2 \theta - \frac{\lambda^2}{3a^2}(1, 3, 4, 7, 9) = \frac{\lambda^2}{4c^2}l^2,$$

then the values of LHS in the ratios of 1, 4, 9, 16 are sorted out and c can be calculated for ratio 1 and using the corresponding values of l as $1, 2, 3, 4, \ldots$, the hk values in the matching ratio are to be found and thus hkl can be indexed.

7.4.4 Orthorhombic System

For this system,

$$\frac{1}{d^2} = \frac{h^2}{a^2} + \frac{k^2}{b^2} + \frac{l^2}{c^2}$$

and applying Bragg's law, we get

$$\sin^2 \theta = \frac{\lambda^2}{4a^2}h^2 + \frac{\lambda^2}{4b^2}k^2 + \frac{\lambda^2}{4c^2}l^2,$$

where $A = \lambda^2/4a^2$, $B = \lambda^2/4b^2$, and $C = \lambda^2/4c^2$ and so, $\sin^2 \theta = Ah^2 + Bk^2 + Cl^2$, the indexing problem is considerably difficult here as three unknowns are to be determined. The general procedure is too lengthy to be discussed here. For example, consider two reflections $hk0$ and $hk1$ say 120 and 121, then the difference of their $\sin^2 \theta$ values gives C and the difference between 310 and 312 is $4C$. However, the difficulty in indexing increases if some lines are found to be missing because of zero structure factor. Moreover, the $\sin^2 \theta$ values are to be determined with considerable accuracy.

7.4.5 Monoclinic and Triclinic Systems

The problem of indexing increases manifold as these systems involve four and six independent constants and there are sometimes more than hundred lines found in the diffraction patterns. Solving and indexing the lines manually then turns out to be almost impossible and so, they are generally done with the aid of computer. Standard programs are now available and therefore indexing of the lines belonging to any system is seldom done manually.

7.5 Summary

1. Different X-ray diffraction techniques for single crystal starting from Laue techniques to rotation/oscillation and finally to Weissenberg techniques are discussed along with the interpretation and the advantages and limitations.
2. These techniques which are systematically introduced and discussed are essential for understanding the inside of structure and their arrangements in a crystal.

References

1. B.D. Cullity, *Elements of X-Ray Diffraction* (Addison Wesley, Reading, MA, 1978)
2. M.J. Buerger, *X-Ray Crystallography* (Wiley, New York, 1942)
3. M.M. Woolfson, *An Introduction to X-Ray Crystallography*, 2nd edn. (Cambridge University Press, Cambridge, 1997)

8

Determination of Space Group and Crystal Structure

8.1 Determination of Space Group from Data Obtained from Moving Film

8.1.1 Weissenberg Photographs

The Weissenberg photographs thus obtained are placed on the Weissenberg chart to find out the coordinates of the diffraction spots in the reciprocal space along with reciprocal cell dimensions and the indices of reflections can also be determined. It may be directly written either on the film or more conveniently on a tracing paper on which the diffraction spots of the Weissenberg have been already traced out. It may be pointed out here that the equi-inclination Weissenberg has some advantage over other settings. The form of the curves for other layer are same as zero layer and the same chart can be used for the interpretation of the photographs. Figure 7.9 is a typical example of zero layer Weissenberg as mentioned and the reason for the formation of festoons is also explained by Fig. 7.10.

After noting down the indices of reflecting planes, the next task is to find whether there are any systematic absences. The absences due to centering of the lattice (f.c.c., b.c.c., etc.) occur throughout the reciprocal space and a general survey will reveal the lattice type which is to be determined. The presence of glide planes and screw axis will affect the absences of certain layers or rows of points in the reciprocal lattice.

To verify and understand how these absences result due to the presence of these microsymmetries, let us consider the following derivation of crystal structure factor in the case of their presence.

The presence of a-glide perpendicular to c-axis means that for any atom at xyz, there must be an identical atom at $x + 1/2, y, -z$ (movement of half of a cell along a which makes x into $x + 1/2$ and reflected across the glide plane which turns z into $-z$). The structure factor will then be $n/2$ of such pair of atoms

$$F = \sum_{n/2} f_r\{\exp 2\pi i(hx_r + ky_r + lz_r) + \exp 2\pi i(h[x_r + 1/2] + ky_r - lz_r)\}$$

$$= \sum_{n/2} f_r \exp 2\pi i(hx_r + ky_r + lz_r)\{1 + \exp \pi i(h - 2)\}. \tag{8.1}$$

Putting $h - 2$ as m, we get $\exp \pi i(h - 2) = \exp m\pi i = \cos m\pi + i \sin m\pi = 1$ or -1 depending on m is even or odd and correspondingly, the reflection will either occur or not.

The situation will be simpler to follow for set of reflections for which $l = 0$, i.e., for zero layer Weissenberg. Equation (8.1) will take a simpler form

$$F = \sum_{n/2} f_r\{\exp 2\pi i(hx_r + ky_r) + \exp 2\pi i(h[x_r + 1/2] + ky_r)\}$$

$$= \sum_{n/2} f_r \exp 2\pi i(hx_r + ky_r)\{1 + \exp \pi i(h)\}.$$

By the same argument, there will be reflections from $hk0$ planes for which h is even and the no reflection when h is odd.

Therefore, reflections will occur only when F_{hk0} does not vanish and it is from the planes $hk0$, where h is even. Similar arguments may also be applied for screw axis of symmetry and other translational symmetries. Screw axis of symmetry produces absences among reflections from the planes perpendicular to them. For example, 4_1 symmetry axis parallel to c limits $00l$ reflections to those with $l = 4n$, while 2_1-axis parallel to a-axis produces reflections $h00$ only when $h = 2n$. These conclusions can be drawn from the calculation of structure factor F_{hkl}.

For example for 2_1 parallel to b, the atoms occur in pairs like x, y, z and $(-x, y + 1/2, -z)$ and the structure factor can be written as

$$F_{hkl} = \sum_{n/2} f\{\exp 2\pi i(hx + ky + lz) + \exp 2\pi i[-hx + k(y + 1/2) - lz]\}$$

$$= \sum_{n/2} f \exp 2\pi i(hx + ky + lz)\{1 + \exp(-4\pi i[hx + lz] + \pi ik)\}.$$

Now, $\exp(-4\pi i[hx + lz] + \pi ik) = \exp -i(m)$, where $m = 4\pi[hx + lz] + \pi k$ and $\exp -i(m) = \cos m - i \sin m = +1$ when m is even and $= -1$ when m is odd. For simplicity, considering reflection $0k0$, we then get $F_{0k0} = 0$ for k is odd and $F_{0k0} \neq 0$ for k is even.

Therefore, whenever this systematic absence is found without others affecting reflections for which both h and l are not zero, then the existence of a 2_1-axis parallel to b is established. Therefore, if systematic absences are found in addition to lattice centering, then it can be concluded that they are due to translational symmetry like glide planes or screw axis present along or perpendicular to the crystal axis.

The list of these systematic absences and the presence of symmetry responsible for it are given in *International Tables* (Vol. 1, pp 111–119 and 133–151 or p 350) which is to be consulted before drawing any conclusion for the Space Group. In the four-digit Space Group symbol, the first digit represents the lattice centering present or not, the second, third, and fourth digits signify the translational symmetries present along or perpendicular to the three crystal axes. This should also be kept in mind that if a, b, and c are interchanged, an apparently different sets of systematic absences will result. For example, the Space Group *Pnam* may be noted as *Pnma* if b and c are exchanged.

Some of the systematic absences due to the presence of translational symmetry elements may be summarized as follows:

If absences are found in reflections like: and the presence of symmetries is:

$0kl$	A glide plane perpendicular to a
$h0l$	A glide plane perpendicular to b
$hk0$	A glide plane perpendicular to c

and if reflections are observed only if: and the presence of symmetries is:

h, k, and l are $2n$	A presence of 2_1, 4_2, or 6_3
and $l = 3n$	A presence of 3_1, (3_2)

The Space Group determination is occasionally complicated by a phenomenon known as *double reflection*. If the beam reflected from a set of planes $h_1k_1l_1$ suffers another reflection at appropriate angle from another set of planes $h_2k_2l_2$, then this doubly reflected beam appears to arise from another set of planes $h_3k_3l_3$, so that

$$h_3k_3l_3 = h_1k_1l_1 + h_2k_2l_2.$$

Though the intensity of the double reflected beam is usually very weak, sometimes both $h_1k_1l_1$ and $h_2k_2l_2$ are very strong giving rise to spurious reflections. However, as this case is not very frequent in the diffraction pattern, the presence of one weak reflection violating the norm of systematic absences may be considered as spurious and can be safely ignored.

8.2 Determination of Crystal Structure

8.2.1 Trial-and-Error Method

It has been stated earlier that the object of crystal structure determination is to find or locate the positions of all the atoms present in the unit cell and thus to determine what it actually means.

The process in general is very elaborate and sometimes very complex when a large number of atoms are involved and it is task to locate their individual positions in space in side the unit cell. Sometimes though now very seldom, it may be a comparatively easier job when some guess can be made about the crystal structure or at least restrict it to much fewer possibilities than many. In the early days of the development of the structure analysis, only simpler types of structure could be tackled and trial-and-error methods based on the special features were used. In trial-and-error method, the structure factors, both calculated F_c and observed F_o, were determined and a factor R known as reliability index or residual defined as

$$R = \frac{\sum ||F_o| - |F_c||}{\sum |F_o|}.$$

Once the trial structure is found which is close to real structure demonstrated by the low value of residual R, the structure can be refined by adopting the routine procedures. This method has now only historical importance as most of the structures that could be solved by this method have already been solved and structure of present-day complex molecules can only be determined by more complex methods. Some important features of these methods are discussed in the following sections in a way which can best be considered as an introduction. The details of these methods can only be found in more advanced books and hundreds of literatures [1–7].

O to O' by a vector $OO' = \mathbf{r}$. Therefore, the Patterson function will be large if the strong regions of electron density are separated by the vector $\mathbf{r}$ *and* if there are several strong regions of electron density separated by a vector $\mathbf{r}$, the $P(\mathbf{r})$ will show a total effect of it and will also be large. The Patterson function will be a superposition of peaks derived from all pairs of atoms in the unit cell. If there is no overlap of Patterson peaks, then the function $P(\mathbf{r})$ will show the position of all interatomic vectors but usually the resolution of Patterson function is very poor and in fact the Patterson peaks are more diffuse than the electron density peaks. Despite this overcrowded nature of the Patterson maps, useful information can be derived and complex structures can be solved from an interpretation of Patterson function.

8.2.2 The Electron Density Equation and Patterson Function

In the expression for the scattering by an atom where there are n number of electrons present as scatterer (the number n is equal to the atomic number of the atom), we get in the amplitude term of the intensity scattered by the atom a factor F as (Appendix A)

$$F(S) = \sum_n \exp(2\pi i/\lambda)(\mathbf{S}' - \mathbf{S}_0) \cdot \mathbf{r_n}$$

$$= \sum_n \exp(2\pi i)\mathbf{S} \cdot \mathbf{r_n}, \quad \text{where} \quad (\mathbf{S}' - \mathbf{S}_0)/\lambda = \mathbf{S}.$$

Now, considering the charges of the electrons as scatterers are to be distributed within a volume and having a charge density $\rho(r)$, we get

$$F(S) = \int_v \rho(r) \exp(2\pi i)\mathbf{S} \cdot \mathbf{r} \, dv.$$

Now, $F(S)$ is the Fourier transform of $\rho(r)$, where

$$\rho(r) = \int_{v^*} F(S) \exp(-2\pi i \, \mathbf{S} \cdot \mathbf{r}) dv^*,$$

the integration is carried out over entire volume v^* of the reciprocal space in which $\mathbf{r}$ is defined.

This electron density is, however, in terms of $F(S)$ which exists in reciprocal space and is a fraction of scattering from an electron considered as a point. Now, for the entire crystal, $F(S)$ has weight at these reciprocal points as $(1/V)F_{hkl}$. Therefore, the equation for $\rho(r)$ can be written as

$$\rho(r) = \frac{1}{v} \sum_h \sum_k \sum_l F_{hkl} \exp\left(-\sum 2\pi i \, \mathbf{S} \cdot \mathbf{r}\right). \tag{8.2}$$

Here, the summations for hkl are from $-\infty$ to ∞.

In this equation, F_{hkl} is the crystal structure factor and $\rho(r)$ is the electron density expressed in electron per unit volume. The structure factor F_{hkl} is given in terms of the fractional coordinates x, y, and z of the atoms in the unit cell and diffraction vector $\mathbf{S}$ in terms of reciprocal vector as

$$\mathbf{r} = x\mathbf{a} + y\mathbf{b} + z\mathbf{c}$$

and

$$\mathbf{S} = h\mathbf{a}^* + k\mathbf{b}^* + l\mathbf{c}^*.$$

Replacing these values of $\mathbf{r}$ and $\mathbf{S}$ in (8.2), we get

$$\rho(xyz) = \frac{1}{V} \sum_h \sum_k \sum_l F_{hkl} \cos\{2\pi(hx + ky + lz)\}. \tag{8.3}$$

Here, again the summations over hkl are from $-\infty$ to ∞.

It has been shown in earlier chapter that the distribution of values of $|F|$ or F^2 gives a measure of intensities and hence the Space Group as different face-centering symmetries result into different distribution of F. Now, as F_{hkl} is complex but the left-hand side of (8.3), i.e., $\rho(xyz)$, is a real quantity, the structure factor F_{hkl} is expressed separately as the summation of one real and complex quantity

$$F_{hkl} = A_{hkl} + iB_{hkl},$$

where

$$A_{hkl} = \sum_{j=1}^{N} f_j \cos\{2\pi(hx_j + ky_j + lz_j)\}$$

and

$$B_{hkl} = \sum_{j=1}^{N} f_j \sin\{2\pi(hx_j + ky_j + lz_j)\}.$$

And also the phase $\tan \phi_{hkl} = B_{hkl}/A_{hkl}$, and now as $F_{hkl} = |F_{hkl}| \exp(i\phi_{hkl})$, we can write (8.3) as

$$\rho(xyz) = \frac{1}{V} \sum_h \sum_k \sum_l |F_{hkl}| \cos\{2\pi(hx_j + ky_j + lz_j) - \phi_{hkl}\}. \qquad (8.4)$$

Now, the problem of determining the crystal structure can be realized. As the structure amplitude $|F_{hkl}|$ can be derived from the measurement of intensity of X-ray reflection, the phase angle ϕ_{hkl} cannot be directly determined and if these phases of the structure factor are known, then the crystal structure is known as one can compute the electron density from (8.4) and hence the positions of the atoms giving rise to the measured electron densities. Therefore, the lack of knowledge of the phases of the structure factors prevents from directly computing the electron density map and hence determines the positions of the atoms. Patterson suggested as an aid the use of the following equation instead of (8.4) [1, 6, 7]:

$$P(\mathbf{r}) = \frac{1}{V} \sum_h |F_h|^2 \exp(-2\pi i\, \mathbf{h} \cdot \mathbf{r}),$$

where $\mathbf{r} = x\mathbf{a} + y\mathbf{b} + z\mathbf{c}$ and $\mathbf{h} = h\mathbf{a}^* + k\mathbf{b}^* + l\mathbf{c}^*$ but as $|F_h|^2 = |F_{\bar{h}}|^2$.

We would get $P(\mathbf{r}) = (1/V) \sum_h |F_h|^2 \cos(2\pi i\, \mathbf{h} \cdot \mathbf{r})$ and so, the Patterson function is real as expected. Now, as the transform of the product of two functions is a convolution of their individual transforms and as the transform of $(1/V)F_h$ is $\rho(r)$ and the transform of $(1/V)F_{\bar{h}}$ is $\rho(-r) = \psi(r)$, say then

$$\int_v \rho(u)\psi(r-u)dv = \frac{1}{V^2} \sum_h F_h \cdot F_{\bar{h}} \exp(-2\pi i\, \mathbf{h} \cdot \mathbf{r})$$

and as $F_h \cdot F_{\bar{h}} = |F_h|^2$, we can write

$$\frac{1}{V} \sum_h |F_h|^2 \exp(-2\pi i\, \mathbf{h} \cdot \mathbf{r}) = V \int_v \rho(u)\psi(r-u)dv = P(\mathbf{r}).$$

We know that $\rho(-r) = \psi(r)$ and so, $\psi(r-u) = \rho(u-r)$ and $P(\mathbf{r}) = P(-\mathbf{r})V \int_v \rho(u)\rho(u-r)dv$ and as $P(\mathbf{r})$ is centrosymmetrical, we get

$P(\mathbf{r}) = P(-\mathbf{r})V = \int_v \rho(u)\rho(u+r)dv$. The physical interpretation of Patterson function may be looked as the superposition of two electron densities of two unit cells whose origins are displaced origins are displaced by r [1].

Sometimes, the solution of structure can be made straightforward if the structure contains a large number of light atoms and a few heavy atoms. The method is then separately named as *heavy atom method* and the structure factor can then be divided into two parts: one due to heavy atom and the other due to light atoms and then it can be written as [1, 6, 7]

$$F_h = C_h + \sum_{j=1}^{N-n} f_j \cos(2\pi \, \mathbf{h} \cdot \mathbf{r}_j) = C_h + K_h,$$

where the contribution due to heavy atoms is C_h and the total effect due to light atoms is K_h. N is the total number of atoms in the unit cell and n is the total number of heavy atoms. Sometimes, the contribution due to these heavy atoms dominates to an extent, so that the most of the F's signs will be same as C_h. This decision will be safe to take if the magnitude Cs are sufficiently high and thus the phase problem can be tackled. It has been observed that the best result is obtained if the average contribution of light atoms and that due to heavy atoms are equal.

The details of the description of the entire range of methods are, however, beyond the scope of this book. It is sufficient for the fulfillment of the aim and ambition of this book if it is well realized that there exists a large collection of formidable armory for the solution of structures even having a very complex nature and only a bird's eye view of some of them is given in this chapter. The details of the processes involved can be found in many books and publications by reputed crystallographers. A short but comprehensive list is given in the chapter for further reading. Therefore, nature and also human synthesize materials and manufacture crystals of them have a three-dimensional pattern and it is also possible to know and determine the symmetry of the entire arrangement of the assembly of atoms in space at determined positions exhibit. It is, therefore, a passage from the symmetry identification to the position determinations of atoms in space and then confirmation of the symmetry that the crystalline structures manifest.

8.3 Summary

A brief discussion along with mathematical basis underlying the different methods applicable for the determination of spatial symmetry present in a crystal, i.e., the Space Group is given. The details of these methods are available in a number of publications and a brief reference is given at the end of the book suitable for Further Reading.

References

1. M.M. Woolfson, *An Introduction to X-Ray Crystallography*, 2nd edn. (Cambridge University Press, Cambridge, 1997)
2. M.J. Buerger, *X-Ray Crystallography* (John Wiley & Sons, New York, 1942)
3. R.W. James, *X-Ray Crystallography* (Methuen, London, 1948)
4. M.J. Buerger, *Crystal Structure Analysis* (Wiley, New York, 1960)
5. H. Lipson, W. Cochran, *The Determination of Crystal Structures* (Bell, London, 1966)
6. Uri Shmuel, *Theories and Techniques of Crystal Structure Determination* (IUCr, Oxford Science, New York, 2007)
7. C. Giacovazzo, *Direct Methods in Crystallography, Fundamentals and Applications* (Oxford University Press, Oxford, 1998)

9

The World of Symmetry

The sky above is adorned with sun and stars
the earth below by full of lives
and that I found for me a place on it,
flabbergasted is aroused, awakened my heart.

_____ **Rabindranath Tagore**

The word "symmetry" constitutes its own world. A world that originates from the natural objects, living or nonliving like from animal bodies or Maple leaves to rock salt or diamond and extends through the man-made objects of great engineering skills like thousands of historically important minarets, modern bridges, sky scrappers, and then through arts like the symmetry in verses and tunes and perhaps finally through science like atomic structure and the characters of physical laws. That is why this world is symmetrically wonderful. Further, as nothing is perfectly perfect, the symmetry what we see in this universe is always associated with asymmetry. Amongst these two extremes it will probably remain as an insolvable question, "which dominates what?" Irrespective of this fact, it may be said that the beauty of the symmetry could probably not realized if it was not intimately combined with asymmetry.

An attempt has been made so far in the earlier chapters to introduce this symmetry present in crystals, which are three dimensional patterns, and the processes and techniques that are followed to know and find this symmetry. Now perhaps it is high time to make an attempt to visualize the symmetry present in nature. Some most common examples are cited only with the intention to get the readers induced if not fully but at least partially so that they would be tempted to peep through the window and have a glimpse of this world of symmetry that lies around us. Now, questions may arise in our mind that what is the domain of this world symmetry? What are its upper and lower limits? If the upper limit is taken as the galaxy our solar system belongs, the lower limit may be taken as the structure of DNA or even lower (Figs. 9.1and 9.2).

Fig. 9.1. The Galaxy, the upper limit? Our own galaxy consists of about 200 billion stars, with our own Sun being a fairly typical specimen. It is a fairly large spiral galaxy and it has three main components: a *disk*, in which the solar system resides, a central *bulge* at the core, and an all encompassing *halo*. The interesting fact that can be observed is its spiral structure

Let us start with the helical spiral symmetry that our Galaxy exhibits in the "Domain of Symmetry." It is interesting to note that both these structures, that is, the Galaxy and DNA show the helical structures. Now within these limits let us examine the symmetry present and we first start with the symmetry present in living bodies.

9.1 Symmetry in Living Bodies

A mirror plane of symmetry actually exists through a vertical plane through any human or animal body (Fig. 9.3) as the right side of the body appear to be just mirror reflection of the left side. A more beautiful and interesting example may be cited as that of the Maple leave or more appropriately the butterfly and snail (Figs. 9.4–9.6). A more complicated symmetry, that is, rotation symmetry is very interestingly manifested in the pattern exhibited by flowers and is given in Fig. 9.7.

Therefore, the symmetry is not only present in the total body of the living objects it goes down even to the molecular level. The living bodies' cells contain chromosome, which carries and preserves the characteristics of the species and also the hereditary information. The chromosomes are consisted

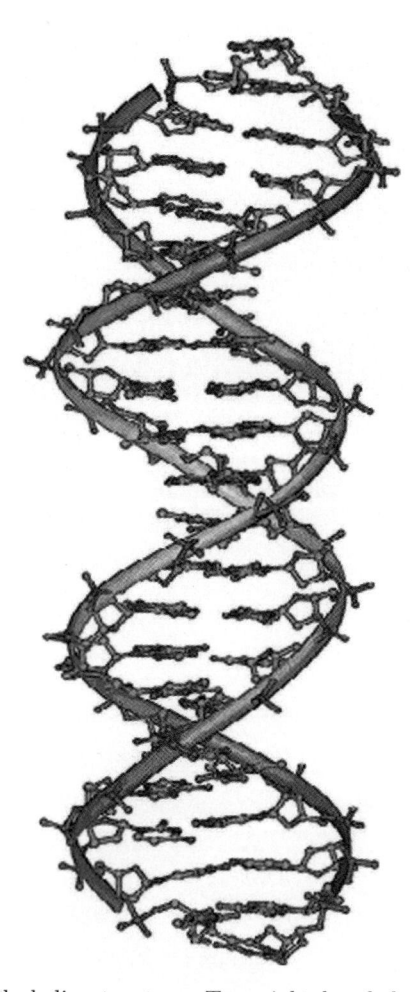

Fig. 9.2. DNA double helix structure: Two right handed screw axes, The lower limit? DNA is made up of subunits, which scientists called nucleotides. Each nucleotide is made up of a sugar, a phosphate, and a base. There are four different bases in a DNA molecule. So, the symmetrical structure of helix exists in the very root of living matter

of proteins and DNA. The genetic as well as the hereditary information are contained in DNA. The structure of this DNA was discovered in the year 1951 by James Watson and Francis Crick based on the X-ray diffraction data obtained by Rosalind Franklin and Maurice Wilkins. The structure of DNA is commonly known as double helix. It is two right-handed screw axis of symmetry about one single axis and structure looks like one twisted ladder. The structure of DNA through its discovery has opened up a fascinating field of

Fig. 9.3. A human figure having a mirror plane of symmetry bisecting vertically the figure

Fig. 9.4. Mirror plane through leaves, that is, Maple leave (*yellow color*) and other leaves and also flowers

Genetic Science and is a certification of the fact the world has symmetry down to the molecule. The work of Watson, Crick, and Wilkins was duly acknowledged through the award of the Nobel Prize in Medicine, and their work was published as "Molecular Structure of Nucleic Acids: A Structure for Deoxyribose Nucleic Acid" in the British journal Nature (April 25, 1953).

Fig. 9.5. A beautiful butterfly, showing a mirror plane of symmetry passing centrally

Fig. 9.6. The logarithmic spiral on snail shell, showing beautiful symmetrical creation of nature under sea. Will that be a too far extrapolation if this spiral nature is compared with that which exists in our Galaxy?

9.2 Symmetry in Patterns, Snow Flakes, and Gems

Man made patterns to paint the walls and floors and even clothes are examples of symmetry of various kinds:

Some "patterns" which apparently look like a pattern but bear no geometric symmetry, yet their looks are pleasant (Figs. 9.8–9.10). Sometimes an asymmetry can also be as beautiful as that one having the conventional symmetry.

While introducing "Crystal," it has been stated earlier that the word has been derived from Greek word meaning ice or frozen water.

In fact a snow flake is a tiny crystal of frozen water and is of various beautiful shapes each having sixfold of rotation symmetry and also a mirror plane of symmetry.

Figure 9.11 gives some examples of snow flake crystals and some gems used in ornaments.

Fig. 9.7. Fivefold of symmetry in flowers. It has been stated and explained earlier that fivefold of symmetry is absent in crystalline materials. It is partially true but it is not true for natural objects like wild flowers; "Hyericum" shows fivefold of symmetry ((**a**) and (**b**)) with or without a mirror plane of symmetry across them

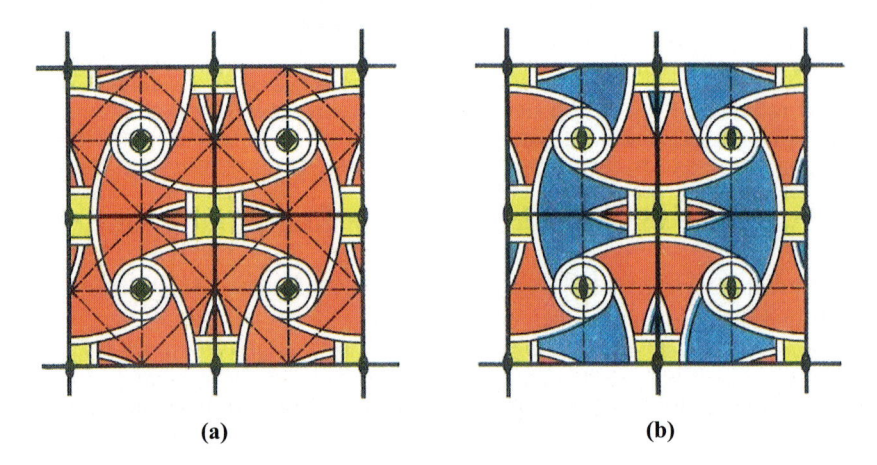

Fig. 9.8. Egyptian pattern. (**a**) There are two mutually perpendicular mirror planes but in Fig. 9.7b, in addition of the same mirror plane, a higher symmetry, that is, fourfold rotation symmetry plus inversion $\bar{4}$ exists perpendicular to the plane of the figure. It should be noted that simply by changing the color of the pattern the symmetry changes or is modified

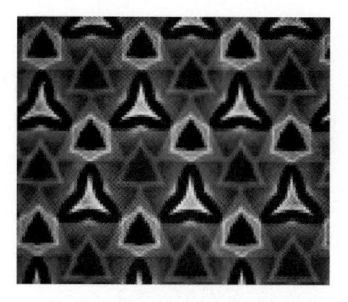

Fig. 9.9. Kaleidoscope pattern. Each motif having threefold of rotation symmetry

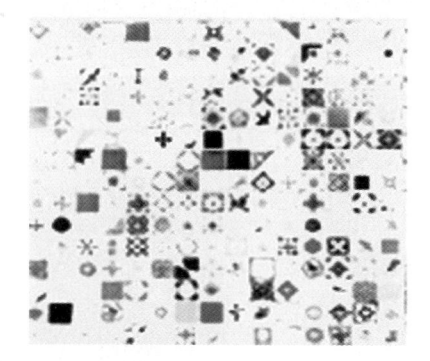

Fig. 9.10. Floor mosaic tiles: No symmetry but is beautiful

9.3 Symmetry in Architecture

In the man-made architectural marvels scattered all over the world, the sense and the knowledge of symmetry that the man acquired has been manifested. To mention and explain a few of them, which have been spread over the span of centuries of the development of human civilization, the Tajmahal of Agra, the pyramids of Egypt, the Eiffel Tower of Paris bear marvelous symmetric structures (Figs. 9.12–9.14). The Cathedral of St. Basil in Moscow is an architectural beauty, which though deviates from geometrical symmetry, yet is an example of asymmetry in symmetry in architecture (Fig. 9.15).

9.4 Symmetry in Fundamental Particles

We have so far seen some natural examples of the presence of symmetry, for example, mirror plane of symmetry in animal bodies, rotation and roto-inversions in precious stones, flowers, and leaves, and now it would be interesting to note that the mirror plane of symmetry is also present in the world of fundamental particles. We know that the atoms are made of positively

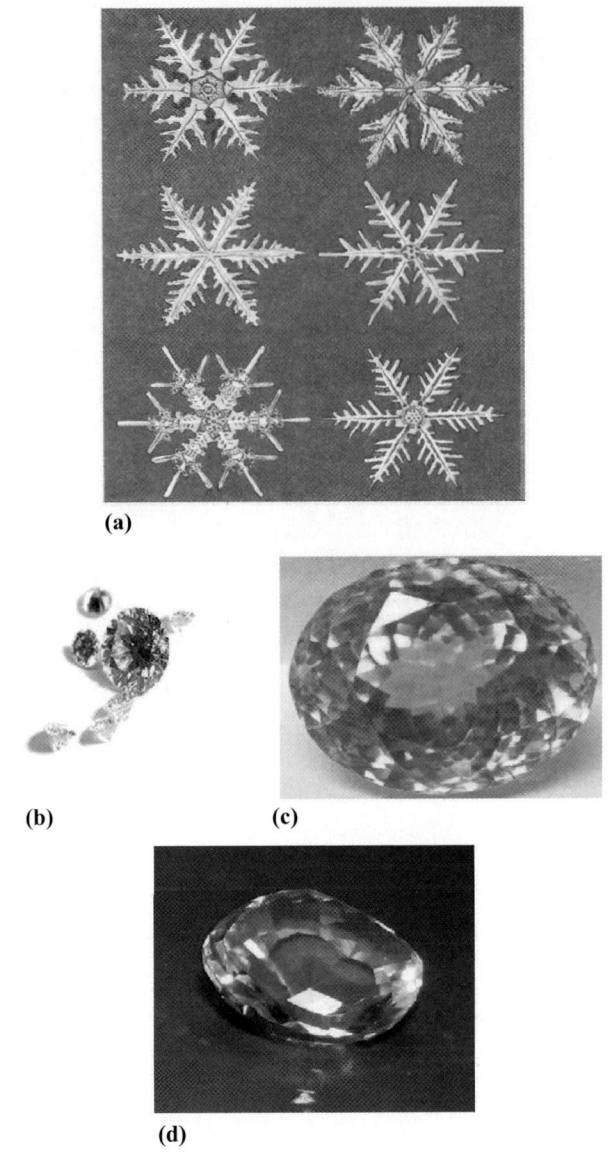

Fig. 9.11. (a) The beautiful six fold structural pattern of snow flakes. (b) Diamond used in ornaments. (c) Ceylon Hessoni. (d) The famous diamond Kohinoor

charged protons and neutral neutrons in the nucleus, the center of atoms, and the negatively charged electrons, which revolve around this center in different orbits. In addition to the interesting fact of finding a symmetry, that is, invariance between this model of atoms and the our solar system having sun at the center and the different planets orbiting round it in different orbits, there

Fig. 9.12. The marvel of one of the man-made symmetries: one of the wonders of the world, the Tajmahal, the tomb of Mughal Emperor Shahjahan and his queen Mumtaz Begum in Agra (built in 1631–1648). The famous architecture has a perfect fourfold of symmetry about the axis vertical through the center of the center dome

Fig. 9.13. The Great Sphinx of Giza with Khafre's pyramid in the background built in the year 2500 BC or perhaps earlier is the symbol of ancient Egyptian culture and an exquisite structure bearing perfect geometrical symmetry

is the existence of some strange particles like antiprotons having mass equal to that of normal protons but having negative charge instead of positive as for protons. The existence of positively charged "electrons" known as positrons has been found, and therefore, the existence of an anti-atom having negatively charged nucleus constituted by antiprotons and positrons in different orbits around this nucleus can not be ruled out. The following list gives some of the particles and their antiparticles. The existence of a mirror plane of symmetry can therefore be imagined between them.

Fig. 9.14. The great Eiffel Tower, Paris (1889), was designed by the famous French engineer Alexandre Gustave Eiffel. Built with 7,000 tons of wrought iron to make the perfect geometrical structure having total height of 320 meters

Objects	Reflected Image ↓	
Mass (Electron Units)	Fundamental Particles	Anti Particles
0	Neutrino ν	Anti Neutrino ν
1	Electron e^-	Positron e^+
207	Mu Mesons μ^-	Mu Meson μ^+
270	Pi Meson π^-	Pi Meson π^+
1837	Neutron n Proton p^+	Anti Proton p^- Anti Neutron n

↑
Mirror Plane

The galaxy constituted by these anti-atoms may exist though the physical verification of their existence may be ruled out as any space ship made from the matter available in our earth will cease to exist as soon as it meets any antimatter of that imaginary antigalaxy. Moreover, it has been theoretically predicted that in our galaxy the existing matter out number the antimatter and that is the reason for the existence of our galaxy and may also be said that the asymmetry exists along with symmetry.

Fig. 9.15. The Cathedral of St. Basilius, Moscow (1551–1561). The architecture by Barma and Postnik and was built by Ivan the Terrible. Architecturally it is a highly successful solution of the problem of symmetry–asymmetry and an elegant exhibition of the artistic solution. It stands as a symbol of esthetic taste and understanding the beauty of the builder

9.5 Symmetry (Invariance) of Physical Laws

The concept of symmetry is not only confined to the symmetry of objects, may it be geometrical patterns or the living or nonliving objects. It is also extended to the physical laws that govern this universe. The German mathematician Hermann Weyl defined symmetry of natural objects as that if something (operation) is done on it, there would be no change on the appearance of the object compared to that before the operation. This is the sense as was narrated by Feynman that the laws of physics are symmetrical. There can be plenty of things that we can do while representing the laws, but this makes no difference and leaves no change in its effect. The symmetry of physical laws resides in their unchangeability or invariance in one or another of transformation, which is related to the change of conditions under which the phenomenon is observed and the law is framed. The invariance, that is, symmetry of the laws of physics under transition from one inertial frame to another inertial reference frame is an example of the symmetry of the physical laws. Any process in nature occurs in the same manner in any reference frame, that is, in all inertial frames a law has the same form. Let us consider for an example the relativity principle [1, 2]

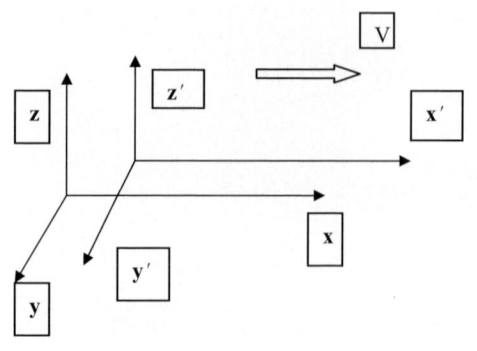

Let x, y, and z be a stationery frame of reference and x', y', and z' be another frame of reference moving with respect to x, y, and z, with a velocity V. Now let an event occur at a time t at a point x, y, z point in the frame x, y, and z. The same event occurs in x', y', and z' at the point x' y' z' at a time t', then the space and time coordinates of the event in frames xyz and $x'\,y'z'$ are related by

$$x' = \frac{x - Vt}{\sqrt{1 - V^2/c^2}},$$

$$y' = y,$$

$$z' = z \text{ and } t' = \frac{t - Vx/c^2}{\sqrt{1 - V^2/c^2}}.$$

This relationship, that is, x' in terms of x are called Lorentz transformations and the symmetry of physical laws with respect to changes from one inertial frame to another are mathematically expressed in this transformation. When the relation is reversed the x in terms of x', y', and z' in terms of y' and z' and t in terms of t' will be given as

$$x = \frac{x' + Vt'}{\sqrt{1 - V^2/c^2}},$$

$$y = y',$$

$$z = z' \text{ and } t = \frac{t' + Vx'/c^2}{\sqrt{1 - V^2/c^2}}.$$

The invariance of the velocity of light c in both of the frames can be shown as:

$$\frac{x'}{t'} = \frac{x - vt}{t - \frac{vx}{c^2}} = \frac{x/t - v}{1 - \frac{vx}{c^2 t}} = \frac{c - v}{1 - v/c} = c.$$

Therefore, if there are two frames of reference moving with respect to each other with constant velocity, then the basic property and value of the physical parameters do not change as for the velocity of light. Now, this Lorentz transformation is simplified in to Galilean transformation when $v \ll c$ as

$$x' = x - vt; y' = y; z' = z, \text{ and } t' = t.$$

So, the laws of physics are symmetrical and invariant when there is uniform motion.

Now for another example, the vector operations, say product of vectors, the replacement of the word "right" by "left," for example, "right hand rule by left hand rule does not make any difference in physical laws.

Let us consider $\mathbf{F} = \mathbf{r} \times \mathbf{p}$, where $\mathbf{F}$ is the torque, $\mathbf{r}$ is the distance of the mass from the axis, and $\mathbf{p}$ is its momentum. From right hand rule of the vector product we get the direction of the product as per the following diagram.

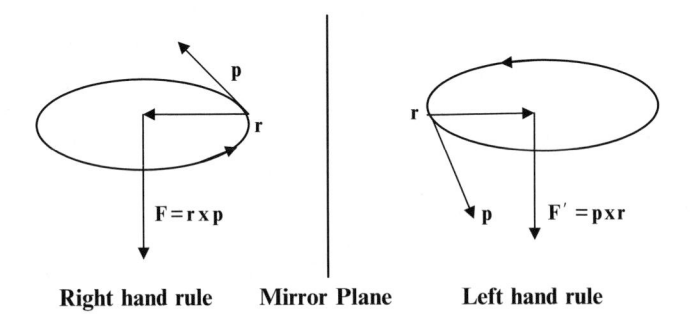

Right hand rule   Mirror Plane   Left hand rule

In the first figure consider a right-handed screw rotated from $\mathbf{r}$ to $\mathbf{p}$, the movement defines the direction of torque, which is down wards. In the second figure imagine a left-handed screw rotating from $\mathbf{p}$ to $\mathbf{r}$, and the direction of torque is invariant as in this unconventional left hand screw the thread is in opposite direction.

The right side of the figure is the mirror reflection of the left side. The directions of $\mathbf{r}$ and that of $\mathbf{p}$ are reversed and now if the right hand rule, which is applied to the left side diagram, is also reversed due to mirror reflection to left hand rule then the direction of torque remains same, which may be thought of as the symmetry or invariance of the physical laws. This may also be seen that if the right hand rule is kept unchanged for the right side diagram, then the direction of the torque will be reversed. The invariance of time was explained by Feynman with an imaginary experiment. If we have two clocks one of which is the exact mirror reflection of the other including the winding of the spring and the motions of the arms, that is, when the arms of the original one move in the "clock wise direction" the other one which is the mirror reflection of the original will have its arms moving in "anti clock wise" direction. Even in this situation it can be easily predicted that both the clocks will "tick" after the same interval and the time in the mirror image clock will not move in the reverse direction. Therefore, the time is invariant even under mirror plane symmetry operation and so there cannot be any "reversibility" of time [1].

Though most of the physical laws bear symmetry, that is, invariance, but it is not always true. The invariance of space can be realized in two frames moving only with uniform motion with respect to each other. But if one of them is moving with acceleration, then an additional force called pseudo force will be acting on the body lying in the accelerating frame of reference, which is now called noninertial frame of reference. Same is the case when we consider two phenomena occurring in two frames when one of which is rotating with respect to the other. The apparatus that measures these phenomena will result differently in the two frames because the apparatus in the rotating one (even with constant angular motion) will experience an additional force known as centrifugal force.

Therefore, there is always some asymmetry in the world of symmetry. Some of the things are not perfectly symmetrical and even if they are symmetrical there is always some gradation of the order of symmetry, that is, from no symmetry to perfect symmetry through the zone of partial symmetry. Sometimes the lack of perfect symmetry at least in some cases is an advantage as because materials having partial symmetry may exhibit some property features, which are special for this state of existence and which the perfectly symmetrical material or phenomena fail to deliver.

The exhibitions of some properties from such partial symmetric or asymmetric materials are described and discussed in the following chapters.

Note: At the end of this chapter a question may linger in the minds of the readers that what precedes what? Is it symmetry that is conceived first and then the objects are tailor made either by man or nature to fit the planned symmetry or is it we who get amazed after finding the symmetry already present in those objects? There lies the controversy. There is probably no specific answer to this question. It is of course more likely that we learned the symmetry present in the nature and tried successfully to emulate them in structures like minarets, sky scrappers, bridges, and planning of cities.

9.6 Summary

1. The symmetrical or the geometrical periodic structures present in living or nonliving bodies are discussed with a few examples.

2. The symmetries in structures are known as in variances in physical laws. It has been established that the physical laws are independent of space and time.
3. It has been also emphasized that the fundamental particles found in this universe may also have their mirror symmetry.

References

1. R.P. Feynman, *Lectures on Physics*, vol 1 (Addison Wesley, New York, 1963)
2. L. Tarasov, *This Amazingly Symmetrical World* (Mir Publications, Moscow, 1986)

10

Asymmetry in Otherwise Symmetrical Matter

We have so far discussed the symmetry present in patterns, crystals, and the physical laws. If symmetry is understood as something that limits the number of possible forms of matter and there can be no existence beyond that boundary defined by this symmetrical world, then it can be found that it is not totally correct. Actually, there is almost no existence of a matter in perfect symmetrical state, and as a consequence, the symmetry must be treated as no more than ideal norm from which there is always deviation in reality. If this deviation is called asymmetry, then the problem of symmetry–asymmetry must be understood more closely and intimately. Symmetry and asymmetry are two closely related phenomena that exist in nature, in substances, and even in physical laws, and in fact they are so closely interlinked that they must be viewed as two aspects of the same concept. If beautiful gems and crystals found in nature are the representation of symmetrical world, water in its bulk structural form shows total asymmetrical arrangement of its molecules.

Therefore, when the aim of the book is to discuss the patterns, crystals, and the symmetry that is manifested by them, it is also necessary to discuss the deviations from symmetry, that is, asymmetry to make it complete or more comprehensive.

10.1 Single Crystals, Poly Crystals, and Asymmetry–Symmetry

The symmetry so far discussed in the earlier chapters consists of several "operations," which when done on the object, the object comes to a stage of self coincidence and there is no difference between the stages before and after the said operations are done. This is due to invariance of these two positions or stages. Now all these symmetries can be regrouped in two broad categories, that is, (1) local order or symmetry and (2) long range order or symmetry depending on the extent of their validity (Fig. 10.1).

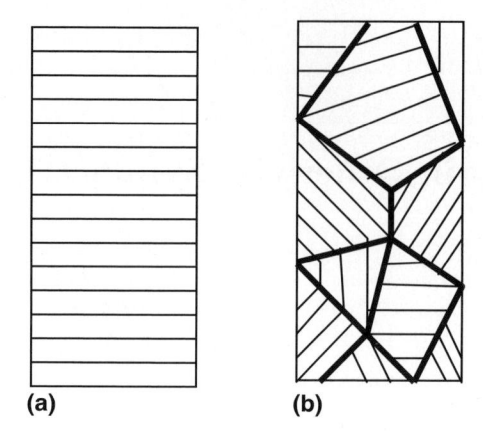

Fig. 10.1. (a) A single crystal having parallel orientation of planes. The alignments of any h_1 k_1 l_1 planes are shown and they are exactly parallel to each other. Both the local and long range orders are valid all through the bulk. **(b)** A polycrystal having random orientation of the same h_1 k_1 l_1 crystal plane. The boundary (*bold lines*) divides the bulk into several regions, separate the regions having same orientation of the plane, and known as Grains. The long range order is assumed to be valid as the disorientation between the planes in the neighboring grains is small but the local order is violated at the grain boundary

This violation of the symmetry on the grain boundary results in a different crystal stage of matter known as polycrystalline state. There is marked difference between the physical properties of these two different stages and many of them, which are characteristic of the respective crystal stage, are important for enhancing the utility of the material. It may be emphasized that the difference between the single crystal and polycrystal state of matter is the randomness of the orientation of any particular plane throughout the bulk. Now this randomness of the orientation will increase if the grain sizes become finer, and this will lead to more asymmetry in one hand and more homogeneity of the physical properties of the material on the other. But there lies enough space in "No man's Land" between these two states of matter. During grain growth state of the heat treatment, the randomness of the orientation of any particular crystal plane will decrease. This is more conveniently achieved by some mechanical processes of deformation. Now, this decrease in the order of randomness of the arrangement of crystallographic planes results in a shift from random orientation to orientation in some preferred direction of the plane. As a result, the homogeneity of physical properties is hampered, giving rise to some heterogeneity in one hand and introduction of some symmetry in otherwise asymmetric stage of matter. This is some times a boon while fabricating some mechanical structure or materials. This phenomena popularly known as "texture" is, however, beyond the scope of discussion of this book and it will suffice if it is mentioned here that this state of material is also a symmetry in the world of asymmetry.

If we summarize the differences between single and polycrystalline states of matter, this may be done as follows:

Property	Single crystalline stage	Polycrystalline stage
Symmetry	Perfect symmetry in ideal crystals	The symmetry is maintained within the region known as grain and remains "almost" same but not exactly same within other neighboring grains belonging to same structure or phase, but is totally different if the grains are of different structures or phases
Order of arrangement of constituents	It remains same both locally and also throughout the bulk	It changes at the boundary between two grains
Physical properties	As the physical properties are direction dependent, a single crystal shows total heterogeneity	A polycrystal in this respect shows homogeneous physical properties

When close-packed structures mainly like FCC and HCP and also other structures are deformed, first thing that happens is the fragmentation of the grains called domains and polycrystalline materials, then shows more homogeneity, and the lattice is strained. This strained lattice contains higher energy and resists more the deformation and thus inducts hardening. There appear some drastic changes in their diffraction patterns. The stacking arrangement of their close-packed planes also changes and this result in a defected region compared to surrounding and is known as stacking fault. The number of planes required to bring back the sequence into original are the number of faulted planes. Less the number of faulted planes for a type of deformation process, more is the energy required. Materials having more of this energy known as stacking fault energy do not usually get "work hardened" (Fig. 10.2).

The displacement of 111 plane by the vector $\mathbf{b} = 1/2[10\bar{1}]$ in one step, that is, from one A site to next A site requires larger misfit energy and so $A \rightarrow B \rightarrow C \rightarrow A$ is preferred by the two partials $\mathbf{b}_1$ and $\mathbf{b}_2$, satisfying the relation $\mathbf{b} = \mathbf{b}_1 + \mathbf{b}_2$, which is

$$a/2\ [10\bar{1}] = a/6\ [\bar{2}11] + a/6\ [\bar{1}\bar{1}2].$$

The above situation may be visualized as follows:

$$\begin{array}{ccccccccc}
A & B & C & A & B & C & A & B & C \\
 & & & \downarrow & \downarrow & \downarrow & \downarrow & \downarrow & \downarrow \\
A & B & C & B & C & A & B & C & A \\
\end{array}$$

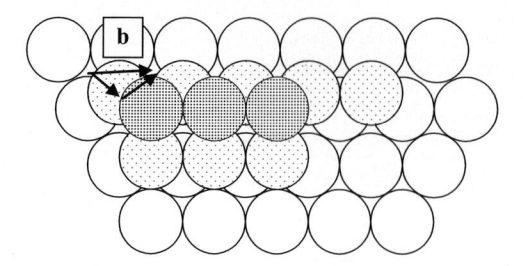

Fig. 10.2. Stacking sequence of close-packed plane {111}. The slip occurs in <110> directions, which involves the sliding of one plane over other. A to B to C by the burger vector b = $1/2[\bar{1}0\bar{1}]$. *Open circles*, A Site; *light-shaded circles*, B Site; *dark-shaded circles*, C Site

↔ Faulted zone having h c p stacking sequence B C B C. type.

Now this faulted region having different stacking sequence does not commensurate with the perfect stacked regions on both sides. This may be seen as an asymmetry introduced in the symmetrical structure.

Introduction of symmetry in otherwise asymmetrical structure is also found in "super lattices" discovered in 1923 in $AuCu_3$ alloys and found later to exist in a number of alloys below a temperature known as critical temperature and they are $PtCu_3$, $FeNi_3$, $MnNi_3$, and (MnFe) Ni_3 alloys. Ordinarily an alloy of say A and B elements exists in solid solutions wherein the atoms of A and B are arranged randomly in the interstitials. This is the state of affairs in the alloys other than those mentioned above. In these alloys, the random structures are available at an elevated temperature, and when they are cooled down below a particular temperature called critical temperature, an ordered state happens wherein a particular set of lattice sites are occupied periodically by say A atoms and the other particular sites by B atoms. The solution is then said to be ordered and the lattice thus constituted is known by super lattice. This is a sort of disorder–order transformation and is manifested by an extra reflection in X-ray diffraction pattern. This is an important phenomena not only because of the fact that this ordered state exhibits different physical and chemical properties, but it is also an example of asymmetry to symmetry transformation. The long range order that exist in the super lattice of $AuCu_3$ alloys can be explained as follows:

In $AuCu_3$ alloys, the occupancy probability for a particular lattice site say by Au atoms is $1/4$, then for Cu atoms it will be $3/4$ because of the composition, and the unit cell for the disordered and ordered structures will look as given in Fig. 10.3.

The view from any side surface of the lattice will demonstrate the super lattice more explicitly.

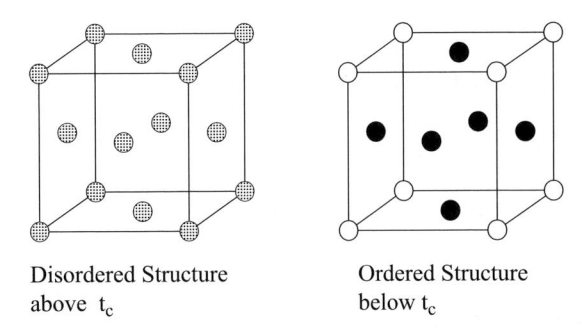

Disordered Structure
above t_c

Ordered Structure
below t_c

Fig. 10.3. Atomic sites having random occupancy by either Au or Cu atoms in disordered structures. *Filled circle*, atomic sites occupied by Cu atoms; *open circle*, atomic sites occupied by Au atoms

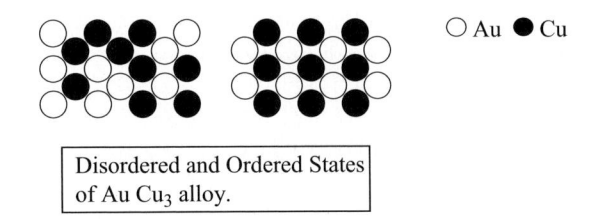

○ Au ● Cu

Disordered and Ordered States
of Au Cu$_3$ alloy.

It can be seen from the above figures that in perfectly ordered state the gold atoms occupy the corner positions and the copper atoms the face-centered positions, whereas in the disordered state there is no such regularity and positions in the unit cell are randomly occupied. As both individual structures are cubic and have almost same lattice parameters, there is only a very slight change in lattice parameter in the ordered state and so there is practically no change in the positions of the X diffraction lines. But the change in the positions of the atoms cause change in the diffracted intensities. Let us see how it changes.

In the disordered state the structure factor F can be calculated as follows:

$f_{av.}$ = (atomic fraction of Au) f_{Au}+ (atomic fraction of Cu) f_{Cu},
$f_{av.} = {}^1\!/_4 f_{Au} + {}^3\!/_3 f_{Cu}$.

The positions of the atoms in the unit cell are 000, $^1\!/_2{}^1\!/_2 0$, $^1\!/_2 0^1\!/_2$, and $0^1\!/_2{}^1\!/_2$.

The structure factor $F = \sum_n f_n e^2 \pi i(hx_n + ky_n + lz_n)$ as there are four atoms in unit cell ($n = 4$).

$$F = f_{av}\left[1 + e^{\pi i(h+k)} + e^{\pi i(h+1)} + e^{\pi i(k+1)}\right]$$

For *hkl* unmixed, $F = 4f_{av.} = (f_{Au} + 3f_{Cu})$ and for *hkl* mixed $F = 0$.

Therefore, the disordered alloy produces the diffraction pattern similar to face-centered cubic structure.

In the ordered state, each unit cell now should contain one Au atom at 000 position and Cu atoms at $1/21/20$, $1/201/2$, and $01/21/2$, and the structure factor then stands out as

$$F = f_{Au} + f_{Cu} \left[e^{\pi i (h+k)} + e^{\pi i (h+1)} + e^{\pi i (k+1)} \right].$$

$F = (f_{Au} + 3f_{Cu})$ when hkl are unmixed, but when hkl are mixed then F instead of being zero it is $F = (f_{Au} - f_{Cu})$. Therefore, so far as the diffraction lines are concerned, there exists one extra line for the reflection hkl mixed (odd and even) only when the structure is perfectly ordered, otherwise it remains as zero. This extra line is the manifestation of ordered structure and is known as "super lattice line" even though they are weaker than fundamental lines.

So, super lattice is a transformation of one "disordered" or asymmetric state to one ordered or symmetric state of this alloy and the order–disorder transformation around a temperature establish that the symmetry and asymmetry may be viewed as the two sides of the same coin.

10.2 A Symmetry in Asymmetry I: Quasi Crystalline State of Matter

It has been discussed in earlier Chapters (Chaps. 3–5) that a perfect crystalline structure should possess a long range order comprising both translational and rotational symmetries, which should be maintained in three dimensions. However, crystalline order can also be maintained in some ways other than translational symmetry and they are called "aperiodic crystals." Now, three alternatives to translational symmetry are known: incommensurately modulated crystals, incommensurate composite crystals, and quasi crystals. The modulated structures are obtained from the structures having translational symmetry by giving displacements of the atoms in the periodic structure by equal amounts. Incommensurate composite structures are formed in layered compounds by two interpenetrating periodic structures which are mutually incommensurate. The discovery of quasi crystals has added up one more dimension to crystallography. Influenced by the discovery of a number of quasi crystals or quasi periodic crystals, International Union of Crystallography has redefined the term crystals to mean "any solid having an essentially discrete diffraction diagram." This broader definition leads to the understanding that microscopic periodicity are sufficient but not necessarily the only condition for crystallinity. A distinct property of quasi crystals that has been found from the diffraction pattern is that it shows fivefold rotation and also other crystallographic point symmetries.

We have seen in Chap. 3 (Fig. 10.3) that there cannot be any crystalline substance with fivefold of symmetry as the motifs having that symmetry cannot make any compact structure, and same is true for sevenfold, eightfold, or tenfold rotation symmetries. It was accepted in classical crystallography that these symmetries are not possible to preserve both translational and rotational symmetries in the long range in stable and metastable states of crystalline solids till the year 1984 [1]. However, quasi crystals as mentioned above lack translational symmetry but rotational symmetries are allowed according to any point group in three-dimensional space. The important logic behind this classical idea was that no compact structure can be formed having fivefold symmetry, but the building principle to form a compact structure can be revised if the motifs are not exactly similar and tilling can be made without overlapping or leaving any gap. This tiling will definitely be aperiodic as they lack translational symmetry and can be taken as a model of quasi crystals. An aperiodic tiling of the plane can be formed with two different proto tiles. In the simplest form, the proto tiles are rhombuses with equal edges but of different angles between the edges. The skinny one has angle $36°$ and the fat one has angle $72°$, that is, a multiple of $(360/10)°$.

Now, not following this matching rule for joining the proto tiles, an infinite number of tilings can be formed, which can be either periodic or aperiodic. One periodic tiling is given above. When this matching rule is followed, the Penrose tiling can be obtained [2].

British mathematician of Oxford, Roger Penrose, devised a pattern in a nonperiodic fashion using two different types of tiles (Fig. 10.4b). The motif of this Penrose tiles is rhombi, which may be arranged in a plane or in three dimension (rhombohedra) so that they obey certain matching rules other than those symmetries discussed before and yet these constitute patterns. Such 2D or 3D tilling have several important properties and among them the most important is that they possess self similarity, which means that any part of the tiling repeats again within a predictable area or volume. This Penrose tiling shows crystalline properties in a number of ways. The edges occur in five different orientations only and thus represent fivefold rotation symmetry. In 1984, when Shechtman et al. [3] published in their paper the electron diffraction pattern of Al-Mn alloy, the diffraction pattern showed tenfold symmetry and that was the first experimental evidence of the presence of symmetry hitherto unpredictable in crystalline matter and it was the evidence of the existence of a new crystalline state knownas quasi crystal. If closely observed there lays a similarity between 3D Penrose pattern and icosahedral quasi crystals. Atomic structures of quasi crystals can be constructed by placing atoms in the Penrose tiling having same atoms at all the vertices and also the similar edges. A three-dimensional structure can then be constructed by stacking the Penrose tilings. This leads to a crystal, which is quasi periodic in two dimension but periodic in third direction. Figure 10.4c, which is the electron diffraction micrograph of rapidly solidified Al-Fe-Cu alloy system distinctly, shows the presence of fivefold of rotation symmetry.

A Fourier transform explains very well the diffraction patterns obtained from Al-Mn quasi crystal. The symmetry that determines the type of quasi crystal is found in its electron diffraction patterns. Figures 10.5–10.7 given below show the diffraction pattern and the simulation of diffraction patterns,

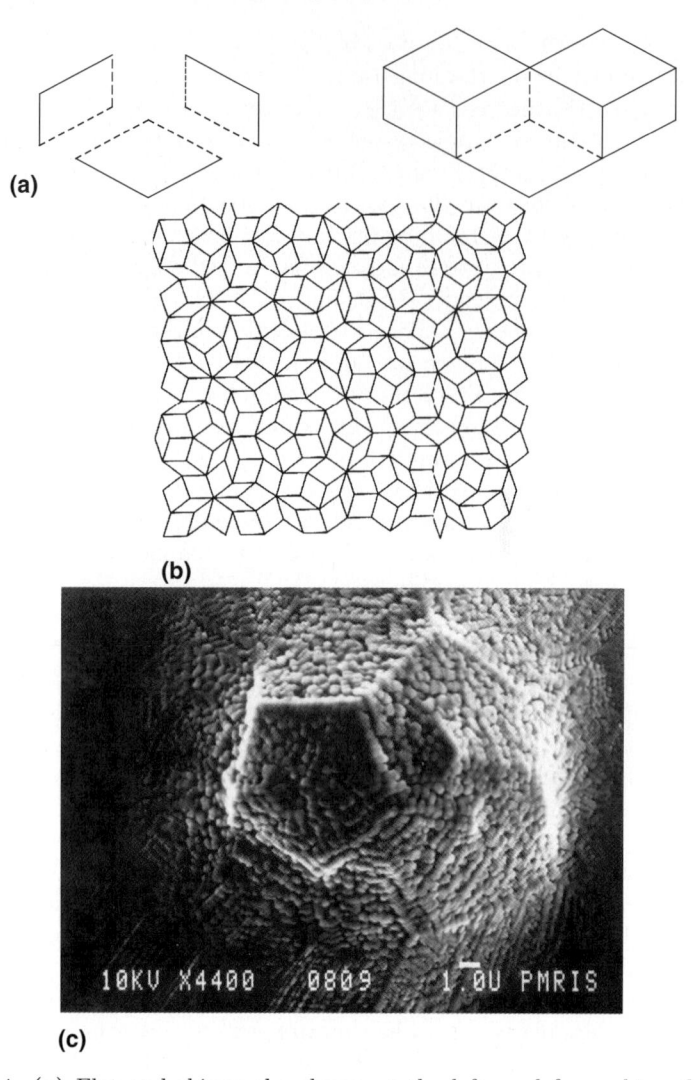

(a)

(b)

(c)

Fig. 10.4. (a) Flat and skinny rhombuses at the left used for making the tiling shown at the right. The sides marked similarly are to be joined to construct the tiling to maintain the matching rule. **(b)** A 2D Penrose tile pattern. A rhombus, that is, motif is arranged in a plane having different modes of arrangements bearing fivefold rotation symmetry and this is repeated in the pattern so that they obey a matching rule. When two of such layers each bearing fivefold symmetry and lying in two planes and rotated by $18°$ with respect to each other so as to constitute a 3D pattern, the projection along the rotation axis results in a tenfold of symmetry. **(c)** The electron diffraction micrograph of rapidly solidified Al-Fe-Cu system. The fivefold of rotation symmetry is shown. (By the courtesy of K. Chattopadhyay, unpublished work)

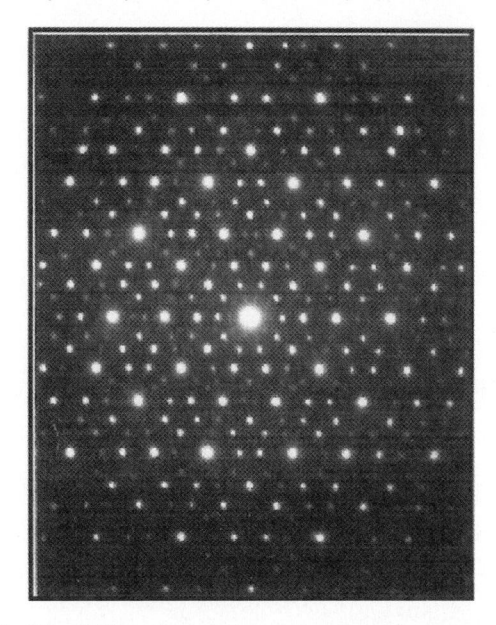

Fig. 10.5. The Diffraction pattern of an icosahedra quasi crystal

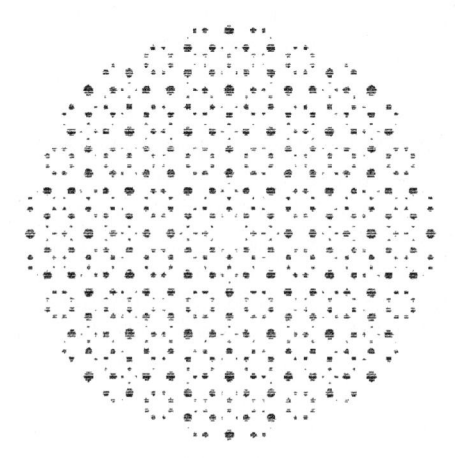

Fig. 10.6. (*Left*) Simulation having eightfold rotation symmetry showing similarity in octagonal quasi crystals

which like others represent the eightfold and tenfold rotation symmetry observed in the electron diffraction or the zero layer precession X-ray photographs.

Since 1984, many stable and also metastable quasi crystals have been found, and these are often binary or ternary intermetallic alloys with aluminum as one of the primary component. Some of these stable quasi crystals with aluminum as a major component are

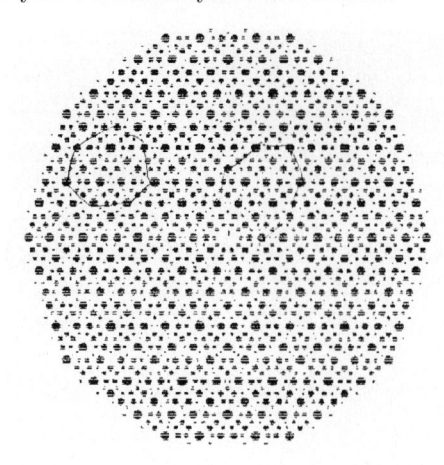

Fig. 10.7. (*Right*) Simulation showing tenfold symmetry decagonal quasi crystals

$$\left.\begin{array}{l} \text{Al-Ni-Co} \\ \text{Al-Cu-Co} \\ \text{Al-Cu-Co-Si} \\ \text{Al-Mn-Pd} \end{array}\right\} \text{Decagonal Quasi Crystals}$$

$$\left.\begin{array}{l} \text{Al-Li-Cu} \\ \text{Al-Pd-Mn} \end{array}\right\} \text{Icosahedral Quasi Crystals}$$

In addition to these stable state quasi crystals, there are much more binary and ternary alloys that form metastable quasi crystalline states.

These quasi crystals are materials with perfect long range order but with no three-dimensional translational periodicity. The first property is manifested by the symmetric diffraction spots and the second property is manifested by the presence of noncrystallographic rotation symmetry, that is, either fivefold or 8-, 10-, or 12-fold rotation symmetry. Since quasi crystals do not show the translational periodicity at least in one dimension, it is mathematically more difficult to interpret its diffraction pattern. As for normal classical crystals, we require three integer values known as Miller indices (*hkl*), to label the observed reflections, because of it having three-dimensional periodicity, for quasi crystals we require at least five linearly independent indices for polygonal quasi crystals and six indices for icosahedral quasi crystals, giving rise to generalized Miller indices. It can then be said that while in three-dimensional space quasi crystals fail to show the required periodicity, in higher dimensional space they exhibit periodicity.

The growth morphology of the stable decagonal quasi crystals Al-Ni-Co or Al-Mn-Pd shows that they grow as decaprismaic (ten prism faces with the tenfold axis as rotation axis). Al-Cu-Fe quasi crystals, which are icosahedral quasi crystals, grow with triacontahedral shape, which exhibit 30 rhombic faces perpendicular to the twofold rotation axes. Though the interpretation of the patterns involves more mathematical complexity, the experimental techniques involved HRTEM (The high resolution transmission electron microscopy) and maximum entropy method (MEM) for the interpretation [4].

Conclusion: A crystalline material should then not be regarded as those materials which only preserve the ten number of macroscopic symmetry in three-dimensional space, but it would be more appropriate to redefine the crystalline materials as those that show a regular diffraction spot and having, in addition to the classically existing ones, also 5, 6, 8, 10 or 12 rotation axes of symmetries.

10.3 A Symmetry in Asymmetry II: Liquid Crystalline Phase

Now, if the long range order (symmetry) is looked in a different way, it can be classified into two categories: (1) positional order and (2) orientation order. This classification is, however, is relevant when the molecules of the "crystalline" matter, the motif of the crystal pattern, are asymmetric in its structure, which is the case in most of the organic and also in some inorganic matter. It is obvious to visualize that if both the categories mentioned above are not valid, then the material cannot posses any crystallinity that is symmetry and it is then called an isotropic liquid stage. But the stage of matter that shows the orientation symmetry but no positional symmetry was discovered in the year 1888 by an Austrian botanist F. Reinitzer while heating a solid compound crystal known as Cholesteryl benzoate. He observed that first the material melt into hazy dense liquid and then on further heating it transformed into a clear liquid. This hazy liquid stage is the new phase of matter and henceforth known as liquid crystal.

Such materials in order to lose the positional symmetry must be liquid having mobility, but its molecules being asymmetric and having the cylindrical shape and being aligned more or less in one particular direction and even being mobile are named as "liquid crystal". Two conventionally diametrically opposite properties are found mingled in this state of matter. It is often called a delicate state of matter. At room temperature if such matter shows liquid crystal characteristic, then upon heating to a high temperature it loses the orientation symmetry and is transformed into an isotropic liquid. If the molecules of this matter are having asymmetric cylindrical shape designated as then in the bulk stage, the three closely comparable stages may be explained by Fig. 10.8a–c.

Therefore, an analogy can be drawn between polycrystals and the liquid crystal. It is being an intermediate state between isotropic liquid and an anisotropic crystal because of having orientation symmetry, while sacrificing totally, the positional symmetry as that of a liquid shows some directionality of the physical properties, more importantly the optical properties. This will be discussed in brief in the later section. Now, it may be mentioned at this

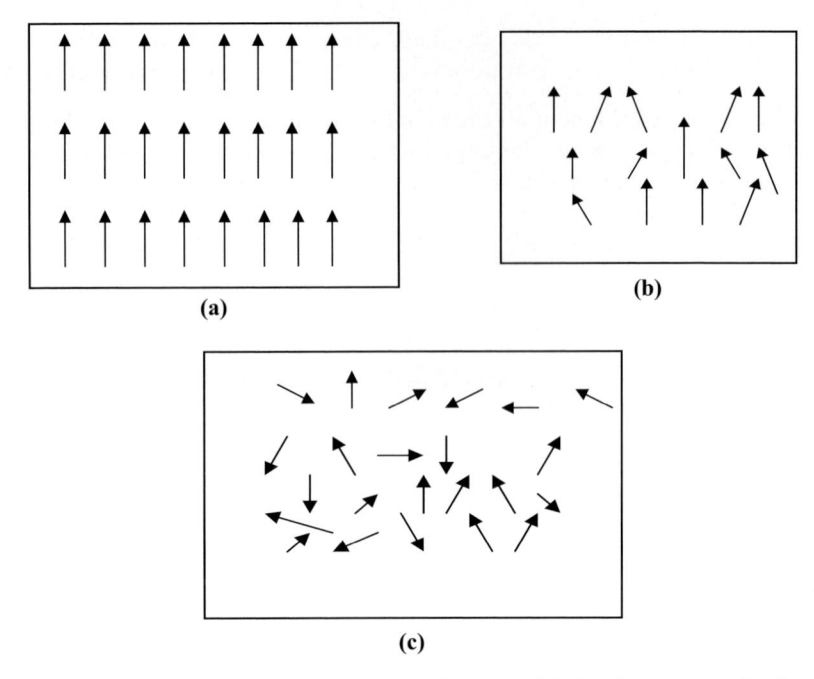

Fig. 10.8. (a) A perfect crystalline state of matter. Molecules preserve both positional and orientation order. **(b)** An intermediate state of matter. Molecules preserve orientation order but no positional order, and the matters preserve fluidity. A state of "anisotropic liquid." **(c)** A totally disordered state of matter, for example, an isotropic liquid. The molecules preserve neither orientation nor positional order

stage that there exists another type of ordering present in materials where the positional order is maintained but not orientation order. That is the state of "Plastic" crystal. Figure 10.9 again classifies these states.

Liquid crystals are of two types:

1. Thermotropic liquid crystals. Where the mesomorphic phase change depends on temperature
2. Lyotropic liquid crystals. Where such phase change depends on the solvent concentration

Thermotropic liquid crystals are classified into the following:

1. Nematic
2. Cholesteric
3. Smectic

Nematic liquid crystals have anisotropic molecules, and they are arranged in space in more or less in one direction called director, but there is no regularity in their positions like liquids. They are only more viscous than ordinary liquids and like liquids they do not possess definite shape. It is the most type of liquid crystal phase [5, 6].

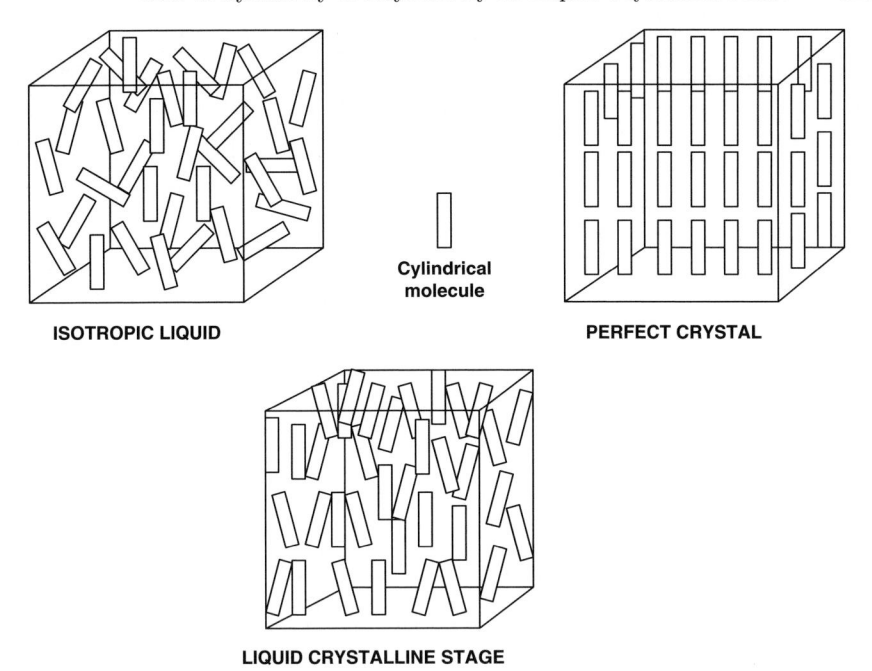

Cylindrical molecule

ISOTROPIC LIQUID

PERFECT CRYSTAL

LIQUID CRYSTALLINE STAGE

Fig. 10.9. The states of isotropic liquids having random arrangement of cylindrical molecules in space, the ordered arrangement (positional and rotational) of the molecules in perfect crystalline stage, and an intermediate stage, that is, liquid crystalline stage

Cholesteric liquid crystals are also nematic, only difference is that their different layers of molecules have helical orientations. The pitch is dependent on temperature. It may also be called as chiral nematic. The nematic liquid crystals may also be made up of "Disc"-shaped molecules instead of cylindrical rod-shaped molecules.

The smectic liquid crystals are of different types, that is, having different arrangement of the anisotropic molecules. Smectic A, Smectic B, Smectic C, and Smectic D (Fig. 10.10). Smectic A is simply a layered structure; there is no order of positional arrangement of molecules in a perpendicular layer. The inter layer attraction is less than intra layer attraction between molecules.

Smectic A type commonly known as simple smectic has nematic structure but having the molecules arranged in layers, and there exists no order of their arrangement in any layer. Smectic B, however, in addition to having the structure as that of Smectic A, the molecules in any layer are arranged bearing sixfold rotation of symmetry. Smectic C has molecules in each layer tilted in more or less in one direction.

The lyotropic liquid crystals exist in the cells of living organism and may be responsible for the proper functioning of the cells, though the discussion on this type of liquid crystal is well beyond the scope of this book, but it

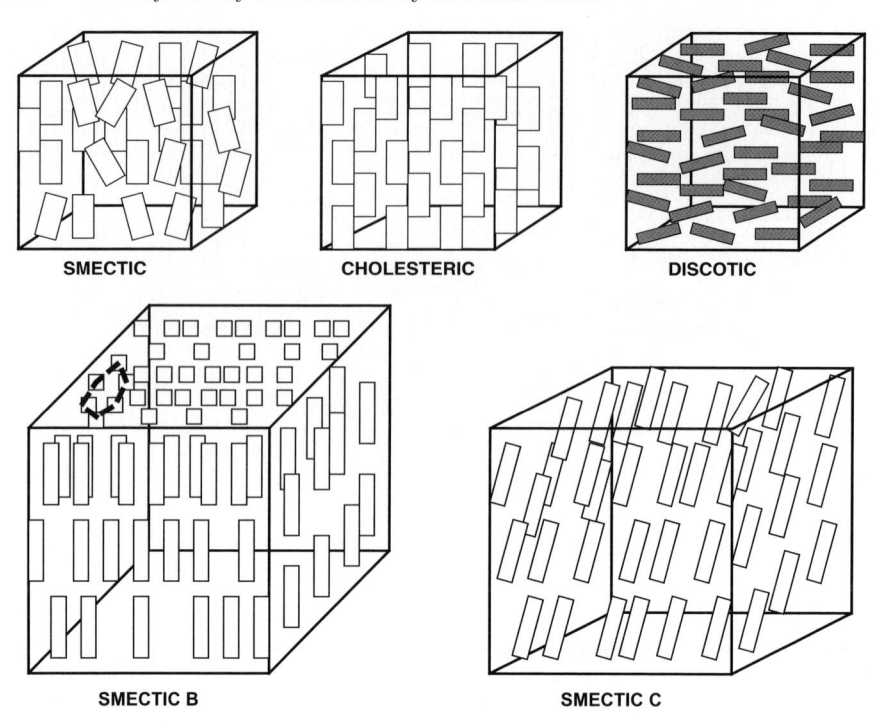

Fig. 10.10. Smectic B and smectic C liquid crystals

may be mentioned here that the structural configuration of these lyotropic liquid crystals present in cells is the essence of life and may be related to the "ordered" state of living matter; any disorder or deviation caused in this constituent may be the reasons behind many diseases [5–8].

The liquid crystal, a delicate state of matter, has enormous applications in display devices. The molecules being mesomorphic in structure and because of the bulk having orientation order, the liquid crystal exhibits optical activity and Birefringence.

10.3.1 Optical Study of Liquid Crystals

The main study of optical properties of liquid crystalline materials is by using a "polarizing light microscope" and it is the first and important property that received attentions (Fig. 10.11).

Birefringence: The difference of velocity of light for different colors and as a consequence different refractive indices is known as dispersion. Because of this the VIBGYOR spectra is exposed whenever white light passes through any refracting medium. Now, the difference in light velocity and the refractive index due to different planes of polarization is known as Birefringence. It is a typical optical property of almost all crystalline substances, which are basically

Polarized Light Microscope Configuration

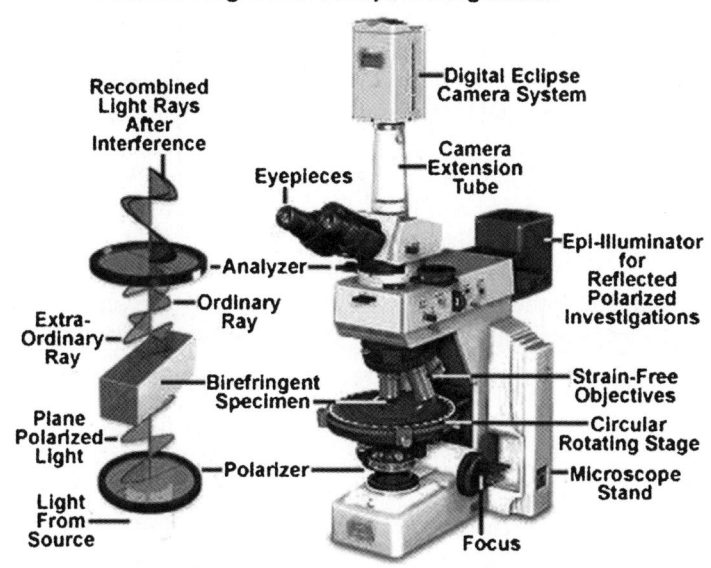

Fig. 10.11. Polarizing optical microscope. (*Right*) The microscope. (*Left*) The path of incident source light through polarizer, birefringent sample (liquid crystals), and analyzer

anisotropic. The anisotropic materials show more than one refractive indices. Light on incidence is split up into two beams; one whose plane of polarization is parallel to the crystallographic axis, that is, optical axis and the other whose polarization plane is perpendicular to the optic axis. These two beams have different velocities of travel through the material and thus have two refractive indices and thus have a resultant phase difference when they go out of the material. When they meet while passing through analyzer, they create an interference pattern depending on the orientation of the molecules of the material. The left side of the above figure shows such effect and the right side shows the polarizing microscope for analyzing this Birefringence in optically active crystals, including liquid crystals. The following figures demonstrate these phenomena in liquid crystal. In a polarizing light microscope, as shown above, the polarizer is set at some angular position and then the analyzer above the specimen stage is rotated until the field of view becomes totally dark, that is, no light can come to the eyepiece. This is known as crossed Nicol position. If an isotropic medium is placed between them on the specimen stage (like glass), the situation does not change. When a liquid crystal, which is Birefringent, is placed on the specimen stage, the situation changes.

Though the details of its various applications are available in some books devoted on liquid crystals and their applications, it may be interesting to introduce the basis of its application in devices that is based on its anisotropic optical property, particularly under electric or magnetic field [9].

10.3.2 Effect of Electric Field on Nematic Liquid Crystal (Electro-Optical Effect)

In Fig. 10.12, the different colored regions in the polarizing light microscope appear due to different orientation of the molecules. The dark regions correspond to the molecules oriented perpendicular to the analyzer and areas of all colored regions are proportional to the number of molecules having same orientation. The molecules of the liquid crystals may be polar molecules or the polarity may be induced by the application of electric field. The application of electric field enhances the orientation in the direction of the field. A transformation from less ordered state to more ordered state starts from more asymmetry to less asymmetry. This can be studied by using one simple specimen stage with electrodes and observing the changes in the birefringence and calculating the areas and their change with the electric field (Fig. 10.13a,b). This can be qualitatively said that while the electric field can enhance the orientation of the molecules in the direction of the field, the other effects like intermolecular attraction and the "anchoring" effect with the glass slide surfaces and the thermal perturbations are the opposing effects. When the areas are plotted with voltage, the resulting curve is shown in Fig. 10.14. The curve shows slow increase of area, particularly of black region, and then the increase becomes rapid and it passes through a plateau region where the effect of field and the opposing effects balance each other. After this region, with increase in field the area increases again. In the Fig. 10.15, the effect is observed with time at a voltage fixed on the plateau region. It also shows a parabolic increase exhibiting the effect of field and the time dependence of the orientation of the molecules [9].

This effect is used in display devices like monitors, watches, and also in other passive matrix and active matrix displays. The total internal effect of the orientation of the plane of polarized light through the birefringent liquid crystals is shown in Fig. 10.16. These devices are efficient power saver and consume on the average 100 times less power than LED.

10.4 Symmetry Down to the Bottom: The Nanostructures

So far we have discussed the different forms of symmetries present in patterns, crystals, and living bodies, but all that were considered exist in bulk where a large number of motifs, atoms, or molecules are involved. Now the question may arise what would happen for the patterns or symmetry if we go down and confine our observations on a very number of motifs? What happens when the total dimension of the pattern is extended only up to a few nanometers or so? The question is typical and the most relevant in the present state of nanomaterials [1].

This small sized pattern (?) may be attained either moving down the dimension level from the bulk or building the bulk from this nanometer level.

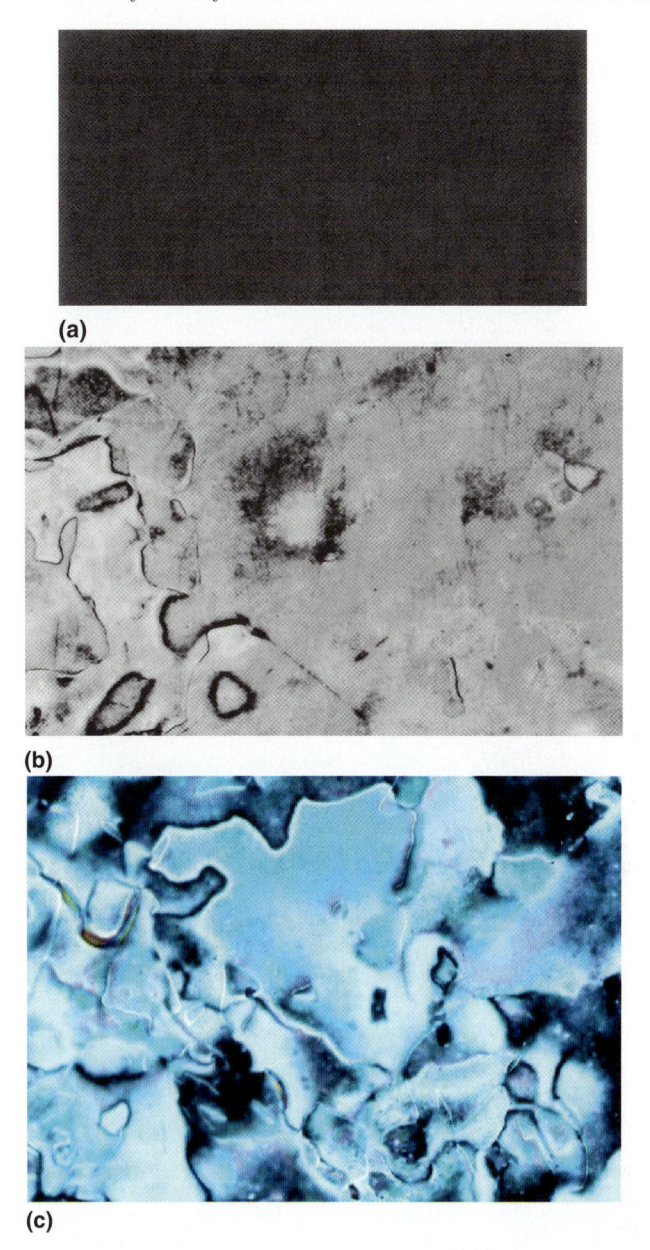

(a)

(b)

(c)

Fig. 10.12. (a) The field in crossed nicol position. Totally dark and no transmission of light through analyzer. **(b)** 4× Magnification birefringent liquid crystal 1-(*trans*-4-Hexylcyclohexyl)-4-isothiocyanatobenzene (Nematic). **(c)** 10× Magnification of the same liquid crystal. The field shows different colored matrices signifying the different orientation of the liquid crystal molecules with respect to the plane of polarized light. The dark regions correspond to the orientation of the molecules perpendicular to the analyzer and white regions parallel to it, whereas the other colors show intermediate orientations

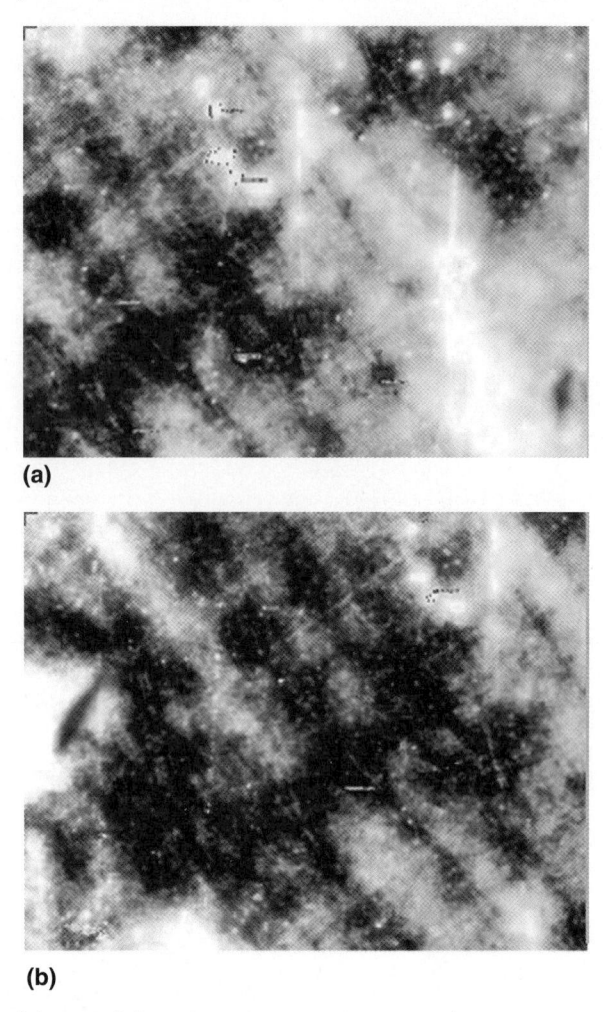

Fig. 10.13. (a) and (b) Electro-optical effect on 1-(*trans*-4-Hexylcyclohexyl)-4-isothiocyanatobenzene (Nomadic). (a) LC sample with no electric field. Areas of black and other colored patches are determined. (b) LC sample with 20 V electric field. Areas of black and other colored patches are determined

The first type of nanomaterials is called the state obtained by "Top Down" and the latter by "Bottom Up." The concept of this top down process was interestingly introduced by Prof. R.P. Feynman in one of his numerous famous lectures titled "There's Plenty of Room at the Bottom," a lecture delivered by him at California Institute of Technology in December 1959. The miniaturization of the bulk was introduced by him in his talk by citing the possibility of writing 24 volumes of Encyclopedia of Britannica on the head of a pin! The technique is well developed now and is known as nano-lithography.

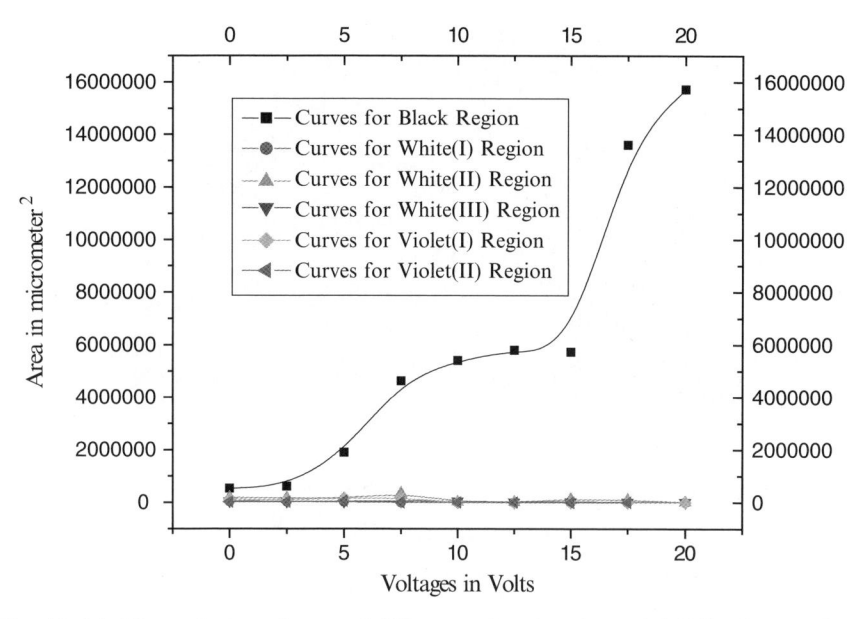

Fig. 10.14. The variation of areas of different colored regions with DC voltage. Black region shows maximum influence of electric field (Fig. 10.13). The plateau is obtained at 10–14 V for 1-(*trans*-4-hexylcyclohexyl)-4-isothiocyanatobenzene (nematic)

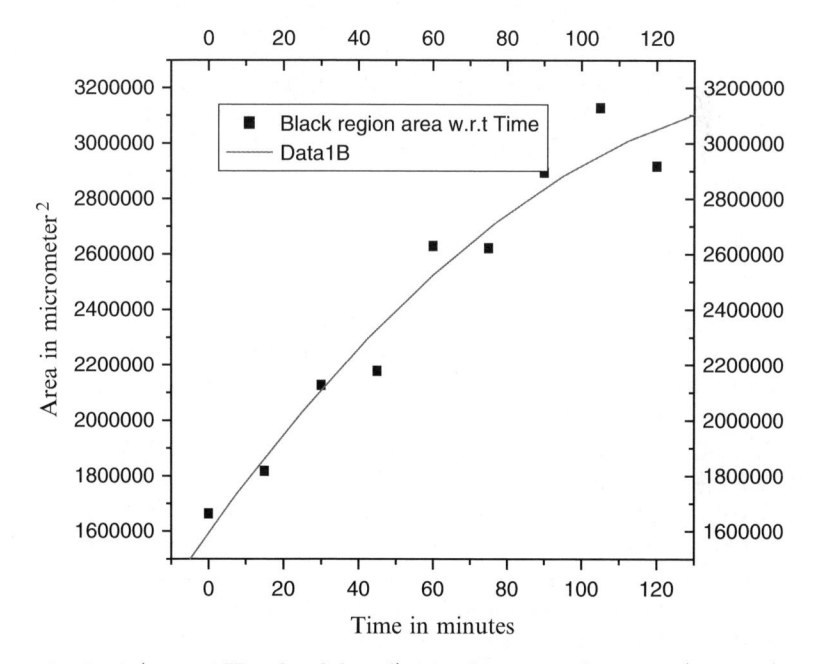

Fig. 10.15. 1-(*trans*-4-Hexylcyclohexyl)-4-isothiocyanatobenzene (nematic). The variation of black area with time at constant voltage of 12 V

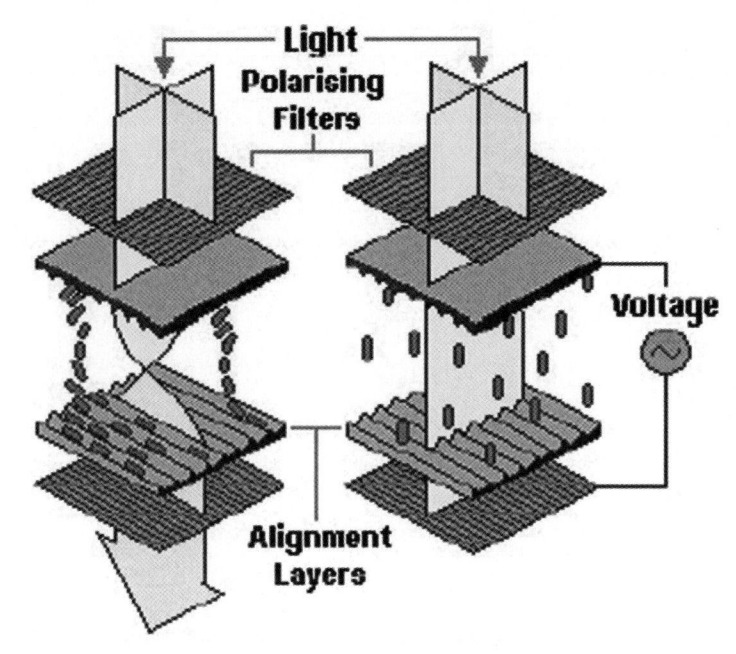

Fig. 10.16. A nematic LC display device. The pink colored plates are two polarizers whose optic axes are perpendicular to each other. The two blue colored plates are glass plates with a transparent plastic and indium tin oxide coating, the latter serve as electrodes. A nematic LC is put in between and the arrangement of its molecules is shown. The first one is the arrangement of a "twisted" stage of molecules under No field condition. The second is the arrangement when there is field. The yellow arrow shows the condition of the polarized light during its transmission when there is no field and its obstruction when the field is on

In his characteristically thought provoking talk he predicted the possibility of miniaturization of computers, circuits, and development of microscopes that will help the Biologists to "see" the mutation of DNA.

The present day production of nanomaterials are through different routs and all of them may be regarded as top down or from bottom up. In one way it is achieved by dividing the bulk materials belonging to one type of structures or arrangement of atoms into small domains of different structures or symmetry of atoms, which are usually scattered in the matrix, and the sizes of these domains are restricted to the size of one to few nanometers. One of these methods is the Ball milling or mechanical alloying. The powders of Co-Ti alloys on ball milling produces amorphous structures, which upon further milling produces nanosized Co_3Ti domains belonging to body-centered cubic and scattered in the amorphous matrix. There are many such examples published in a number of publications, showing the precipitation of one phase (structure) of nanosize in the matrix of other different structure. The methods of the production of nanostructured materials can be broadly classified into three categories:

I. Physical Methods: Mechanical alloying (Ball Milling), Vapor deposition, Sputtering deposition, chemical vapor deposition, etc.
II. Chemical Methods: Colloidal route, Sol–gel route, etc.
III. Biological Methods: Protein synthesis, Synthesis of microorganism, etc.

The detailed discussion of these methods is however beyond the scope of this book.

The bottom up methods result in tailor-made materials. We can arrange without violating the physical laws layers of few atoms in three dimensions so that in none of the direction the dimension increase beyond the range between 1 and 100 nm. We can make something marvelous and that is what exactly we can do. The materials of this nanoscale dimension are not only governed by the quantum mechanical laws as the dimension is so small that the classical mechanics does not hold good, they show some different properties both physical, optical, mechanical, and also the structural. We know that when any colored glass is broken into pieces, its color does not change, but when it is grounded into powder, then irrespective of its original color it turns into white. This is due to light scattering from the fine particles of the powder. When it is nanoparticles its color changes through different shades as its particle size changes. For example, when we consider CdS nanoparticles, its color systematically changes from light pink to deep reddish pink when the particle size changes from 2 to 6 nm. It becomes light colored when particle size decreases. In essence, they completely belong to a different symmetry than the bulk state or conventional state of this same system.

In this bottom up or auto assembly method, the individual molecules have their edges so encoded that they automatically connect each other in the correct manner.

One of the most talked about state is the third state of carbon other than graphite and diamond. It is called fullerene structure of carbon (Fig. 10.17).

Carbon 60 (C60, Buckyball) is this third form of carbon, discovered in 1985 by Richard Smalley, Harold Kroto, and Robert Curl for which they won the 1996 Nobel Prize in chemistry. It is named as "Buckministerfuller" to honor the architect of the geodesic dome, Buckminster Fuller, because the dome's shell resembles the fullerenes' hollow-core construction. Fullerene structure of carbon is face-centered cubic having carbon molecules at the corners and at the center of the faces and belonging to the fullerene family. In the world of symmetry it is definitely a new form of pattern created by the existing symmetry operations.

These tiny tubes of carbon, crafted into the shape of a Y, could revolutionize the computer industry, as they act as remarkably efficient electronic transistors – the toggles used to control the flow of electrons through computer circuits (Fig. 10.18). But the nanotransistors are just a few hundred millionths of a meter in size, which is roughly 100 times smaller than the

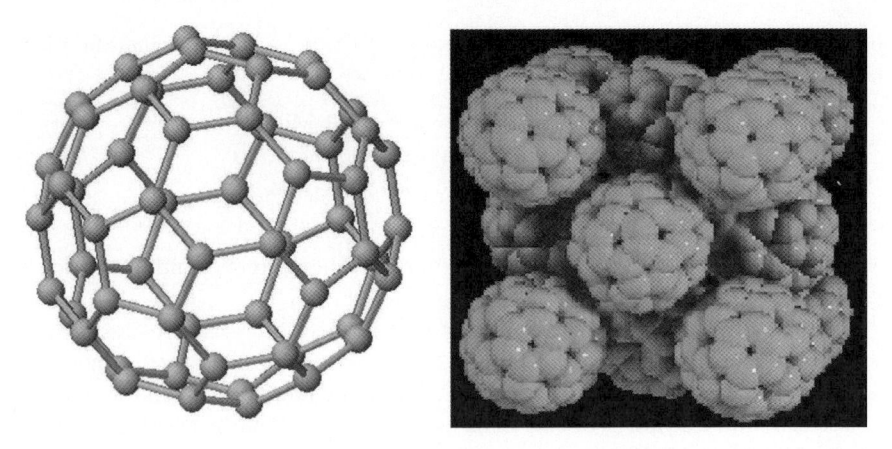

Fig. 10.17. Left is the fullerene structure of C 60 and right shows the face-centered unit cell with fullerene structured carbon atoms at the eight corners and at the center of the six faces

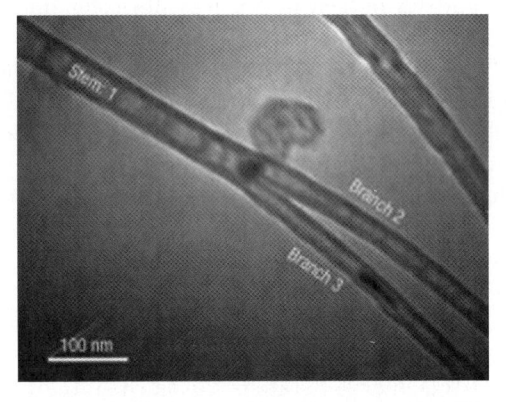

Fig. 10.18. Y-shaped nanotubes are readymade transistors. New research suggests that tiny tubes of carbon crafted into the shape of a Y could revolutionize the computer industry. The work has shown that Y-shaped carbon nanotubes are easily made and act as remarkably efficient electronic transistors – the toggles used to control the flow of electrons through computer circuits. But the nanotransistors are just a few hundred millionths of a meter in size – roughly 100 times smaller than the components used in today's microprocessors. They could, therefore, be used to create microchips several orders of magnitude more powerful than the ones used in computers today, with no increase in chip size

components used in today's microprocessors. They could, therefore, be used to create microchips several orders of magnitude more powerful than the ones used in computers today, with no increase in chip size. Prab Bandaru and his colleagues at the University of California in San Diego, and Apparao Rao of Clemson University in South Carolina, both in the US, started by growing ordinary carbon nanotubes through chemical vapor deposition. Iron–titanium particles were added to spur the growth of an extra nanotube branch attached to the main stem. The overall structure assumed a Y-shape and the catalyst particles were absorbed into the tubes at the branching point (see Fig. 10.10).

Experiments showed that applying a voltage to the stem of the Y precisely controls the flow of electrons through the other two branches. The switching capacity of these nanostructures is in comparable to that of today's silicon transistors. It is true that the present day silicon transistors have been shrunk to around 100 nm, but this Y-shaped nanotubes measure just tens of nanometres in size. Eventually, they could even be shrunk to just a few nanometers.

"The transistor is fully self-contained, the discovery heralds a new era of nanoelectronics in that functionality can be harnessed using all-carbon devices." says Bandaru.

The next generation of computer and television screens could be built using carbon nanotubes. A prototype high-definition 10-cm flat screen has been already made using this technology.

The new screen, called a nano emissive display or NED, is made from two sheets of glass, one covered by a layer of nanotubes standing on end, the other by a layer of blue, red, or green phosphors similar to those used in conventional cathode ray tube screens. When charged, the nanotubes direct electrons at the phosphors, making them light up. Because the electrons have only a short distance to travel, even a 105-cm NED would use relatively little power, says maker Motorola. A screen that size will also have a wide viewing angle and could sell for less than \$400, the company claims (Excerpts from New Scientist magazine, 21 May 2005).

Further a material that is harder than diamond has been created in the lab by packing together tiny "nanorods" of carbon. This new material, known as aggregated carbon nanorods (ACNR), was created by compressing and heating super-strong carbon molecules called buckyballs or carbon-60 consisting of 60 atoms that interlock in hexagonal or pentagonal shapes and resemble tiny soccer balls (Fig. 10.9). This super-tough ACNR was created by compressing carbon-60 to 200 times normal atmospheric pressure, while simultaneously heating it to $2226°C$. The properties of the resulting material were then measured using a diamond anvil cell. This instrument squeezes a material between two normal diamonds, enabling researchers to study it at high pressure using synchrotron radiation – extremely intense X-rays which reveal the material's structure. The researchers found their ACNR to be 0.3% denser than ordinary diamond and more resistant to pressure than any other known material and can actually scratch a normal diamond.

While an ordinary diamond gets its hardness from the strong molecular bonds between each of its atoms, ACNR derives its strength from the fact that it is formed by interlocking nanorods.

Finally, the future of these nanoscale materials, their production processes, and characterization has ushered in a new era. This symmetrical arrangement of motifs down to the bottom has tremendous application potentialities, which are radically different from their bulk counter part and probably a minute part of its applicability in industry and particularly in life sciences has yet been explored and established.

"The precise and energy-efficient self-assembly of matter into material structures with properties that cannot be achieved otherwise is an important goal for nanotechnology," said Mihail Roco, NSF senior advisor for nanotechnology and chair of the National Science and Technology Council's Subcommittee on Nanoscale Science and Engineering. "This is just one way that nanotechnology will help foster "the next industrial revolution."

There are different microscopes to observe the nanostructure. The grazing angle X-ray diffraction, Low Energy Electron diffraction (LEED), Electron Tunneling microscopy, and Scanning electron microscopy are amongst the conventional experimental techniques for studying the surface morphology of the nanomaterials (Fig. 10.19).

Another important instrument to study the nanostructure is the atomic force microscope (AFM) (Fig. 10.20). An atomic force microscope is optimized for measuring surface features that are extremely small, thus it is important to be familiar with the dimensions of the features being measured. An atomic force microscope is capable of imaging features as small as a carbon atom and as large as the cross section of a human hair. A carbon atom is approximately .25 nanometers (nm) in diameter and the diameter of a human hair is approximately 80 microns (μm) in diameter.

Traditional microscopes have only one measure of resolution; the resolution in the plane of an image. An atomic force microscope has two measures of

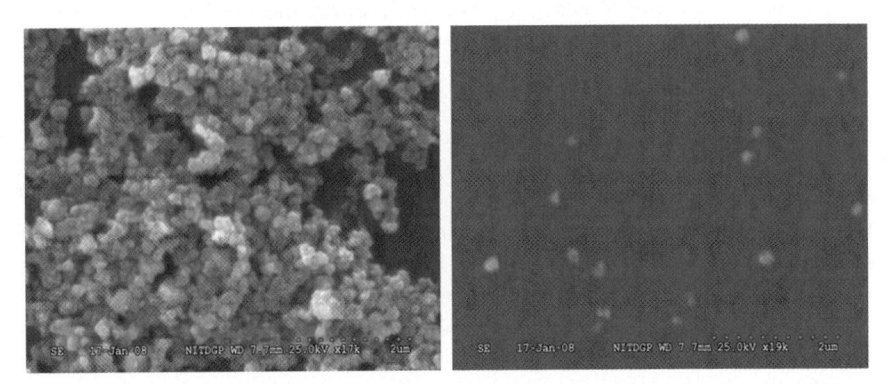

Fig. 10.19. Scanning electron micrograph of nanocrystalline Nickel ferrite sample prepared by chemical route with ball − milling time of 1 h. The picture indicates that the grains are well resolved and have almost spherical in shape. The particle sizes are varying between 20 and 50 nm. Figure on the left indicates that the small particles are getting agglomerated and the figure on the right indicates the single particle picture. By courtesy of S. Ghatak and A.K. Meikap (unpublished work)

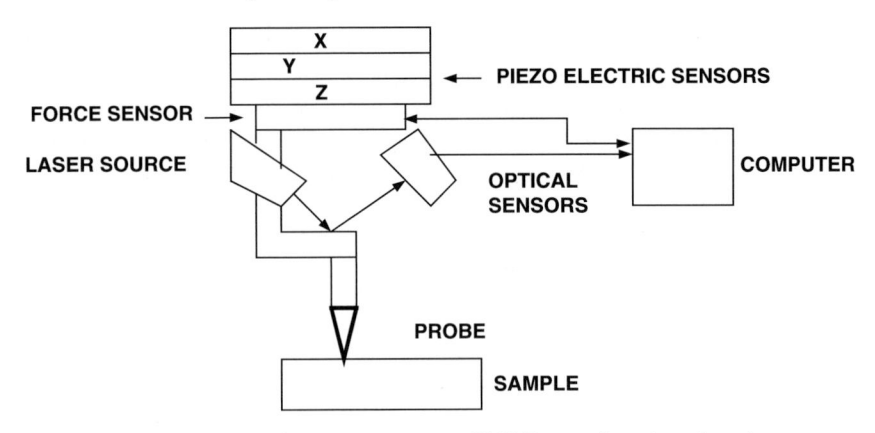

Fig. 10.20. The atomic force microscope. X,Y,Z are the piezoelectric sensors to control the motion of the probe over the sample in XYZ directions. Force sensor is to measure the force between the probe and the sample. (CPU) Computer – The computer is used for setting the scanning parameters such as scan size, scan speed, feedback control response, and visualizing images captured with the microscope

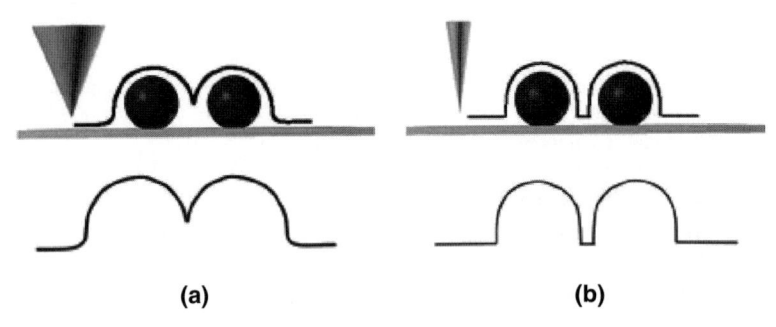

Fig. 10.21. (a) A less sharp probe giving less resolution of the atoms. The atoms appear to be overlapping. (b) A sharper probe than (a). Figure shows more resolved positions of atoms

resolution; in the plane and perpendicular on the sample surface. The planar resolution depends on the geometry of the probe that is used for scanning. In general, the sharper the probe is, the higher the resolution of the AFM image (Fig. 10.21).

Therefore, as a conclusion it may be said that whether the object is in the bulk state or in the nanostate of matter, there exists an order which is followed for the growth of the material from bottom to top and it continues to exist when we go down from top to bottom. The order may change during this transition but as a whole it exists in most of the cases and these results in too many drastic changes in the properties of the material.

10.5 Summary

Along with symmetrical structures or the periodic structures, there is as a rule the existence of asymmetric or aperiodic structures. It has also been introduced that materials present in their asymmetric structural existence show drastically different characteristics, which are not found to exist in their periodic structural states of existence. This asymmetries, which are present in the aperiodic structures of materials and which are incommensurate with the classical structures, leads to the revelation of many more fantastic properties, which are hitherto not found in perfect structural existence of the same materials.

References

1. A. Inoue, K. Hashimoto, *Amorphous and Nanocrystalline Materials* (Springer, Heidelberg, 2001)
2. R. Penrose, Bull. Inst. Math. Appl. **10**, 266 (1974)
3. D. Shechtman, I. Blech, D. Gratias, J.W. Cahn, Phys. Rev. Lett. **53**, 1951 (1984)
4. S.M. Allen, E.L. Thomas, *Structure of Materials, MIT Series*, (Wiley, New York, 1999), Ch. 4, p. 213
5. M.R. Fisch, *Liquid Crystals, Laptops and Life* (World Scientific, Singapore, 2004)
6. P.J. Collings, *Liquid Crystals, Nature's Delicate State of Matter* (Princeton University Press, Princeton, 1990)
7. B. Bahadur (ed.), *Liquid Crystals, Applications and Uses*, vol.1–3 (World Scientific Publishing, Singapore, 1990)
8. S. Chandrasekhar, *Liquid Crystals*, 2nd edn. (Cambridge UniversityPress, Cambridge, 1992)
9. L.M. Blinov, V.G. Chigrinov, *Electro Optic Effects in Liquid Crystal Materials*, (Springer-Verlag, Berlin, 1996)

Epilogue

The present endeavour in the form of a book is a brief sojourn in the fascinating world of symmetry that dwells in patterns, crystals and the physical laws and that which decors and governs this universe. It is true that in the short life-span of human beings, the symmetries in objects, incidence and the phenomena that happen to exist in front of our eyes every day attract the attention and absorb the minds of inquisitive individuals. The vastness of the ocean, the expanse of the meadows, the wonderful symmetry of blooming flowers and their petals of innumerable colours and the endless starry sky have been inciting a quest in the mind to know this wonderful world we live in.

It should be recorded that from antiquity to the present, the notion of symmetry has undergone a lengthy development. From a purely geometrical concept it turned in to a fundamental notion lying at the foundation of nature's laws. Now, we understand that the symmetry is not only that which is visible to our eyes. The symmetry is not just the thing that exists around us but it is the root of everything. According to the modern views, the concept of symmetry is characterized by a certain structure in which three factors are combined:

(1) An object (phenomenon) whose symmetry is considered, (2) transformations or the operations under which the symmetries are considered and (3) the invariance (unchangeability or the conservation) of some properties of the object that expresses the symmetries under consideration. Invariance exists not in general but only in as far as something is conserved.

Symmetry in essence limits the number of possible forms of natural structures and also the number of possible forms of behaviour of various systems.

It can be said that there are three stages of the cognition of our world. At the lowest stage is the phenomena, at the second upper, the laws of nature and lastly at the top third stage symmetry principles. The laws of nature govern phenomena, and the principle of symmetry governs the laws of nature. If the natural laws enable the phenomena to be predicted, then the symmetry principle enables the laws of nature to be dictated. There are also the laws and the concept of compensation of symmetry: once the symmetry is lowered at one level it is conserved at the other at a larger level. It is directly related to the problem of symmetry–asymmetry. We know how strongly today's picture of the physically symmetrical world is different from the geometrically symmetrical cosmos of the ancients. The symmetry must be treated as no more than ideal norm from which there are always deviations in reality. Symmetrical crystal changes to the partially symmetrical liquid crystals and the conventional geometrical symmetry in crystals to the unconventional symmetry in quasi-crystalline materials. Thus, the problem of symmetry–asymmetry must be understood more deeply. Symmetry and asymmetry are so closely interlinked that they must be viewed as the two aspects of the same concept. In a less complex manner this proposition can be viewed as, that the beauty of a face sometimes increases if it has a dimple on only one side of the face than symmetrically on both, a clear deviation from symmetry yet having greater visual pleasure. So said the Soviet philosopher V. Gott in his book *Symmetry and Asymmetry* as: "Symmetry discloses its content and meaning through asymmetry, which in itself is a result of changes, or violations of symmetry. Symmetry and asymmetry is one of the manifestations of the general law of dialectics – the law of unity and conflict of opposites."

More we grasp the symmetry of nature, more asymmetry comes out. Therefore, any search for a unified theory or universal equations is bound to fail as it tantamount to an attempt to consider symmetry separately from asymmetry. The two dialectically opposite categories, symmetry and asymmetry can not exist independently. In an absolutely symmetrical world we would observe nothing – no objects, no phenomena. A crystal otherwise symmetrical can only be grown on the basis of a linear defect i.e. Screw dislocation.

In exactly the same way, an absolutely asymmetrical world is impossible to find too. Behind almost every discovery and its attempts, the driving force was the sense of symmetry and the efforts to establish that. We can site numerous examples like Maxwell's displacement current to wave particle dualism and also to De Broglie's finding of electron wave.

We see, therefore, that symmetry is dominant not only in the process of scientific quest but also in the process of its sensual, emotional perception of the world. In Nature–science–art we find the age old competition of symmetry and asymmetry and this competition is the leading force between all changes and transformations.

A

An Outline of the Diffraction Theory of Periodic and Aperiodic Structures

The detailed theory of X-ray diffraction from periodic and aperiodic structures is, however, out of the scope of this book. However, a brief introduction is given here. The readers are requested to find the detailed analysis from [1–4] and also they may consult the references given in "Further Reading."

Scattering by an Isolated Electron

When an unpolarized X-ray beam is incident on an electron, the electron will experience a force due to the electric field vector of the X-ray electromagnetic wave. Due to this force, the electron will accelerate and as an accelerated charge radiates, electron will also radiate the energy and this is known as *scattering by an electron*. Assuming that there is no net absorption of energy by the electron, there will be no change of wavelength of the scattered X-ray. Though this elastic scattering is an approximation, it is close to the actual. The following figure demonstrates such scattering.

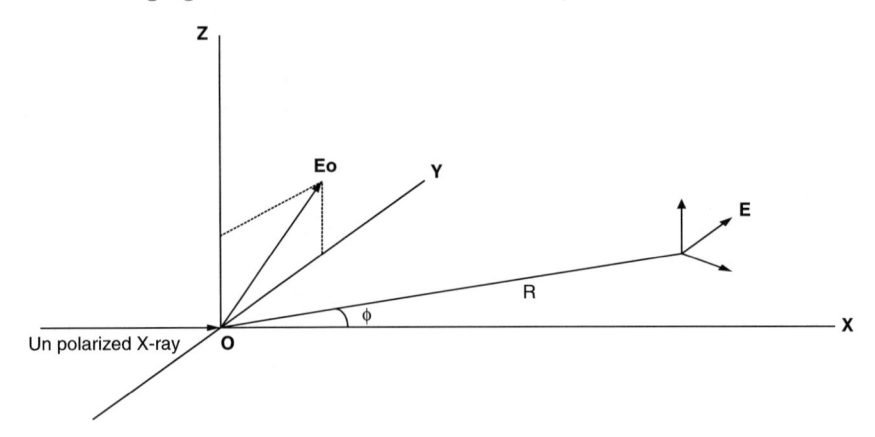

The force experienced by an electron of charge e due to a field ε having an instantaneous amplitude on the YZ-plane as E_0 will be given in Y and Z

directions as $F_Y = E_{OY} e \sin 2\nu t$ and $F_Z = E_{OZ} e \sin 2\nu t$. The accelerations resulted in Y and Z directions are

$$f_Y = \frac{E_{OY} \sin 2\pi\nu t}{m} \quad \text{and} \quad f_Z = \frac{E_{OZ} \sin 2\pi\nu t}{m}.$$

The field scattered by the accelerated electron is

$$\frac{\varepsilon}{Y} = \frac{e^2 E_{OY} \sin 2\pi\nu t \cos\varphi}{mc^2 R}$$

as the direction of the scattered field due to E_{OY} component includes the sin of the angle between Y and R, which is $90° - \varphi$. But the Z component is perpendicular to the XY-plane on which R lies, the field scattered by the Z component is

$$\frac{\varepsilon}{Z} = \frac{e^2 E_{OZ} \sin 2\pi\nu t}{mc^2 R}.$$

The time averages on the square of the amplitudes when added give the resultant amplitude of the scattered wave and that is given by

$$\langle E^2 \rangle = \langle E_0^2 \rangle \frac{e^4}{m^2 c^2 R^2} \left(\frac{1 + \cos^2\varphi}{2} \right).$$

And in terms of intensity, the intensity of the scattered radiation at a distance R from origin is given as

$$I = I_0 \frac{e^4}{m^2 c^2 R^2} \left(\frac{1 + \cos^2\varphi}{2} \right).$$

This is known as classical Thomson Scattering equation.

Diffraction by Periodic Perfect Small Crystal

The electromagnetic radiations like X-rays are scattered by electron cloud around the center of the atom. This cloud of electrons will have a charge distribution resulting into a charge density ρ and the charge in an elementary volume element dv due to a single electron will be ρdv, so that $\int \rho dv = 1$ in terms of electron units. Figure A.1 demonstrates such scattering.

The X-ray field scattered from this spherically symmetrical electron charge distribution (per electron) and received at the point P is given by

$$\varepsilon_P = \frac{E_0 e^2}{mc^2 R} \cos \left[2\pi\nu t - \frac{2\pi}{\lambda}(X_1 + X_2) \right]. \tag{A.1}$$

Now, $X_1 + X_2 = \mathbf{r_n} \cdot \mathbf{S_0} + R - \mathbf{r_n} \cdot \mathbf{S} = R - (\mathbf{S} - \mathbf{S_0}) \cdot \mathbf{r_n}$ and so, the above expression can be modified for the electron cloud (per electron) in terms of complex exponential as

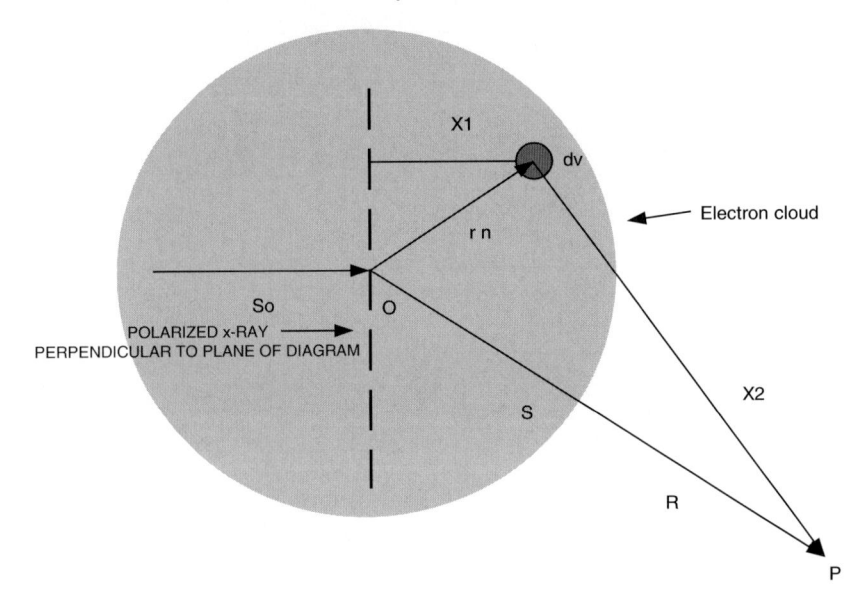

Fig. A.1. $\mathbf{S}_0$ and $\mathbf{S}$ are directions of incident (plane polarized) and diffracted X-ray from the volume element $\mathrm{d}v$ of the electron cloud. $\mathbf{r_n}$ is the position vector of volume element $\mathrm{d}v$ having charge $\rho\,\mathrm{d}v$. X_1 is the distance of the volume element $\mathrm{d}v$ from the plane-polarized X-ray wave and X_2 is the distance of the point of observation from the volume element $\mathrm{d}v$

$$\varepsilon = \frac{E_0 e^2}{mc^2 R}\exp 2\pi\mathrm{i}[vt - R/\lambda]\int \exp[(2\pi\mathrm{i}/\lambda)(\mathbf{S} - \mathbf{S}_0)\cdot\mathbf{r}]\rho\,\mathrm{d}v. \qquad (A.2)$$

Now, $(\mathbf{S}-\mathbf{S}_0)\cdot\mathbf{r}$ can be expressed as $2\sin\theta r\cos\varphi$ and expressing $\mathrm{d}v$ in spherical coordinate as the charge distribution bears a spherical symmetry, we get $\mathrm{d}v = \mathrm{d}x\cdot\mathrm{d}y\cdot\mathrm{d}z = \mathrm{d}r\cdot r\,\mathrm{d}\varphi\cdot r\,\sin\varphi\,\mathrm{d}\psi$. Integrating over ψ from 0 to 2π, we get $\mathrm{d}v = 2\pi r^2\sin\varphi\,\mathrm{d}\varphi\mathrm{d}r$.

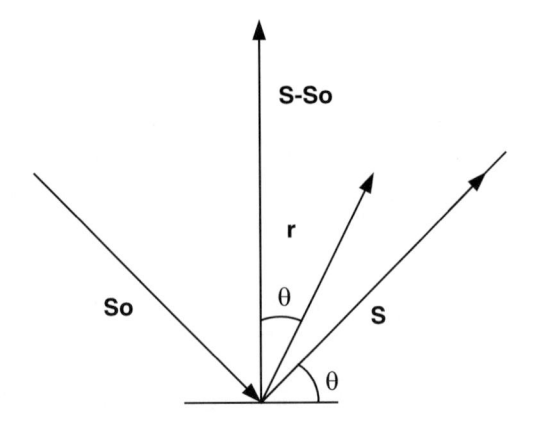

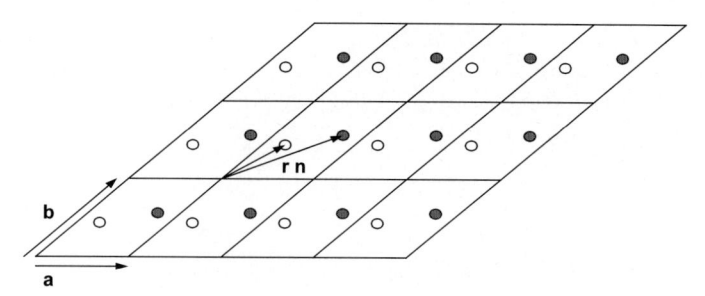

Fig. A.2. A section of a two-dimensional crystal

Now, replacing $\mathrm{d}v$ and writing ρ the charge density as function distance r from 0 and abbreviating $4\pi \sin\theta/\lambda = k$, we write (A.2) as

$$f_e = \int_{r=0}^{\infty} \int_{\phi=0}^{\pi} \exp(ikr\cos\phi)\rho(r)2\pi r^2 \sin\phi \; \mathrm{d}\phi\mathrm{d}r.$$

Integrating over φ, we get

$$f_e = \int_0^{\infty} 4\pi r^2 \rho(r)\frac{\sin kr}{kr}\mathrm{d}r.$$

As this is applicable for only one single electron and there are n such electrons in one atom, so writing f for atomic scattering factor, we get

$$f = \sum_n \int_0^{\infty} 4\pi r^2 \rho(r)\frac{\sin kr}{kr}\mathrm{d}r. \tag{A.3}$$

A perfect crystal is made up of unit cells periodically arranged. Figure A.2 shows a two-dimensional representation of perfect crystal and unit cell parameters are designated as $\mathbf{a}$ and $\mathbf{b}$. The position vector in terms of fractional coordinates x_n, y_n, and z_n of an nth atom is given by $\mathbf{r_n}$ in any unit cell and it may be given in a three-dimensional crystal by $\mathbf{r_n} = x_n\mathbf{a} + y_n\mathbf{b} + z_n\mathbf{c}$ and the position vector with respect to crystal origin is given by $\mathbf{R_{m}n} = m_1\mathbf{a} + m_2\mathbf{b} + m_3\mathbf{c} + \mathbf{r_n}$, where $\mathbf{a}$, $\mathbf{b}$, and $\mathbf{c}$ are the unit cell vectors and m_1, m_2, and m_3 are the integers specifying the position of the unit cell with respect to the origin. For simplicity, let unit vectors in direct space henceforth be written as $\mathbf{a}_1$, $\mathbf{a}_2$, and $\mathbf{a}_3$ in place of $\mathbf{a}$, $\mathbf{b}$, and $\mathbf{c}$ as noted earlier.

The instantaneous field scattered by a small crystal at a distance R from the crystal is then given by

$$\varepsilon = \frac{E_0 e^2}{mc^2 R} \exp[2\pi i(\upsilon t - R/\lambda)] \sum_n f_n \exp[(2\pi i/\lambda)(\mathbf{S} - \mathbf{S}_0) \cdot \mathbf{r_n}]$$

$$\times \sum_{mi,ai}^{i=1,2,3} \exp[(2\pi i/\lambda)(\mathbf{S} - \mathbf{S}_0)m_i a_i], \tag{A.4}$$

where $\sum_n f_n \exp[(2\pi i/\lambda)(\mathbf{S} - \mathbf{S}_0) \cdot \mathbf{r_n}]$, over all such n number of atoms in a unit cell, is known as "the crystal structure factor" and henceforth will be noted as F. Now, any factor of the summations over $m(= 1, 2, 3)$ and $a_{1,2,3}$ leads to three Laue equations as

$$(\pi/\lambda)(\mathbf{S} - \mathbf{S}_0) \cdot \mathbf{a}_1 = h\pi, \quad \text{i.e.,} \quad (\mathbf{S} - \mathbf{S}_0) \cdot \mathbf{a}_1 = h\lambda,$$
$$(\pi/\lambda)(\mathbf{S} - \mathbf{S}_0) \cdot \mathbf{a}_2 = k\pi, \quad \text{i.e.,} \quad (\mathbf{S} - \mathbf{S}_0) \cdot \mathbf{a}_2 = k\lambda,$$
$$(\pi/\lambda)(\mathbf{S} - \mathbf{S}_0) \cdot \mathbf{a}_3 = l\pi, \quad \text{i.e.,} \quad (\mathbf{S} - \mathbf{S}_0) \cdot \mathbf{a}_3 = l\lambda,$$

where h, k, and l are integers and are the Miller indices of the crystal planes. These three equations must be simultaneously satisfied so as to result any diffraction or reflection from the plane.

Introducing the reciprocal lattice vectors as $\mathbf{b}_1$, $\mathbf{b}_2$, $\mathbf{b}_3$ in place of $\mathbf{a}^*$, $\mathbf{b}^*$, $\mathbf{c}^*$ and a vector in reciprocal space as $\mathbf{H}_{hkl} = h\mathbf{b}_1 + k\mathbf{b}_2 + l\mathbf{b}_3$, it can be shown that the diffraction vector $(\mathbf{S} - \mathbf{S}_0)$ in reciprocal space can be related to the reciprocal lattice vector $\mathbf{H}_{hkl}$ as

$$(\mathbf{S} - \mathbf{S}_0) = \lambda \mathbf{H}_{hkl},$$

which is the Bragg's law and $|\mathbf{H}_{hkl}| = 1/d_{hkl}$. Now abbreviating $(\mathbf{S} - \mathbf{S}_0)/\lambda$ as $\mathbf{S}$ and $m_1\mathbf{a}_1 + m_2\mathbf{a}_2 + m_3\mathbf{a}_3$ as $\mathbf{L}$, we get

$$\mathbf{R_{mn}} = \mathbf{L} + \mathbf{r_n}.$$

The time-independent amplitude term of the scattered field ε can be written after using the abbreviations as above

$$E(\mathbf{S}) = F(\mathbf{S}) \sum_{L}^{N} \exp(2\pi i \, \mathbf{S} \cdot L),$$

where N is the number of such unit cells in the diffracted volume of the crystal. In terms of reciprocal lattice vectors $\mathbf{H}$ and using the delta function $\delta(\mathbf{S})$ which is defined by $\delta(x - a) = 0$ except at $x = a$ and also for well-behaved function as

$$\int_{-\infty}^{+\infty} f(x)\delta(x - a)\mathrm{d}x = f(a)$$

and also using the property

$$\sum_{m=-\infty}^{\infty} \exp(2\pi i \, \mathbf{S} \cdot m) = \sum_{m=-\infty}^{\infty} \delta(\mathbf{S} - m), \tag{A.5}$$

we can write

$$E(\mathbf{S}) = F(\mathbf{S}) \sum_{H} \delta(\mathbf{S} - H). \tag{A.6}$$

Diffraction by Modulated Structures

Modulated structures can be obtained from the structures having translational symmetry by say a displacement of atomic layers by an integral number of lattice translations. This results into a superlattice still retaining translational symmetry. But if the displacement is not an integral multiple of lattice vectors, the resulting lattice will lose its commensurability with the basic structure and also the translational symmetry. The diffraction of the modulated wave in addition to the determination of $\mathbf{r_n}$ of the atoms is within the domain of "Incommensurate Crystallography."

The scattered X-ray wave from such modulated structures is in the form of Bragg's reflection and the scattering vectors of Bragg's reflections are

$$\mathbf{H'} = h\mathbf{b}_1 + k\mathbf{b}_2 + l\mathbf{b}_3 + m\mathbf{b}_4,$$

where $\mathbf{H'} = \mathbf{H} + m\mathbf{b}_4$ and h, k, l, and m are the four reflection indices or $\mathbf{H'} = \mathbf{H} + m\mathbf{q}$, where $\mathbf{q} = \mathbf{b}_4$ and have the significance of being a wave vector giving the modulation functions. The diffraction pattern of such modulated structures will contain Bragg's reflection of the basic structure given by $\mathbf{H}$ (i.e., $m = 0$) and they are found to be surrounded by satellite reflections corresponding to having $m \neq 0$.

Now, the difference in these modulated structures from the periodic counterpart starts with the position of the atoms, which in modulated structure are given by

$$\mathbf{r'_n} = \mathbf{L} + \mathbf{r_n} + \mathbf{u^n}(\overline{x4}), \qquad (A.7)$$

where $\mathbf{u^n}(\overline{x4}) = u_1^n (x_4)\mathbf{a}_1 + u_2^n (x_4)\mathbf{a}_2 + u_3^n (x_4)\mathbf{a}_3$ gives the displacement of atom out of its basic structural position and $n = 1, 2, \ldots, N$ the atoms in the unit cell of the basic structure, and $\mathbf{L}$ is the lattice vectors of the basic structure. Using these relations, the amplitude of the scattered X-ray $E(\mathbf{S})$ will be given by

$$E(\mathbf{S}) = \sum_{L}^{N_{\text{cell}}} \sum_{n}^{N} f_n(\mathbf{S}) \exp[2\pi i \, \mathbf{S} \cdot (\mathbf{L} + \mathbf{r_n} + \mathbf{u^n}(\overline{x4})], \qquad (A.8)$$

where $\mathbf{S}$ as defined earlier is $(\mathbf{S} - \mathbf{S}_0)/\lambda$ and f_n is the atomic scattering factor of nth atom. As $E(\mathbf{S})$ is a complex function multiplying it with its complex conjugate, we can get the expression for the intensity of the scattered radiation from modulated structures having the modulation incommensurate with the basic structure.

References

1. B.E. Warren, *X-Ray Diffraction* (Addison-Wesley, Reading, MA, 1969)
2. M.M. Woolfson, *An Introduction to X-Ray Crystallography* (Cambridge University Press, Cambridge, 1970)
3. S.K. Chatterjee, *X-Ray Diffraction: Its Theory and Applications* (Prentice-Hall, New Delhi, 1999)
4. S.V. Smaalen, *Incommensurate Crystallography* (IUCr, Oxford Science, New York, 2007)

B

Solved Problems

1. Let $\mathbf{a}$, $\mathbf{b}$, $\mathbf{c}$ and $\mathbf{a}^*$, $\mathbf{b}^*$, $\mathbf{c}^*$ are direct lattice and reciprocal lattice vectors, respectively. Now, V_D and V_R are, respectively, the volumes of unit cells in direct and reciprocal lattices and are given as

$$V_D = \mathbf{a} \cdot \mathbf{b} \times \mathbf{c} \quad \text{and} \quad V_R = \mathbf{a}^* \cdot \mathbf{b}^* \times \mathbf{c}^*.$$

Show that $V_D = 1/V_R$.
Solution:
We know that

$$\mathbf{a}^* = \frac{\mathbf{b} \times \mathbf{c}}{V_D}, \quad \mathbf{b}^* = \frac{\mathbf{c} \times \mathbf{a}}{V_D}, \quad \text{and} \quad \mathbf{c}^* = \frac{\mathbf{a} \times \mathbf{b}}{V_D},$$

$$V_R = \mathbf{a}^* \cdot \mathbf{b}^* \times \mathbf{c}^* = \frac{(\mathbf{b} \times \mathbf{c}) \cdot [(\mathbf{c} \times \mathbf{a}) \times (\mathbf{a} \times \mathbf{b})]}{V_D^3}$$

$$= \frac{(\mathbf{a} \times \mathbf{b}) \cdot (\mathbf{b} \times \mathbf{c}) \times (\mathbf{c} \times \mathbf{a})}{V_D^3}$$

$$= \frac{(\mathbf{a} \cdot \mathbf{b} \times \mathbf{c})^2}{V_D^3} = \frac{V_D^2}{V_D^3} = 1/V_D.$$

Therefore, $V_D = 1/V_R$.

2. Two crystallographic directions are given as $\mathbf{A}$ and $\mathbf{A}'$, and they are

$$\mathbf{A} = u\mathbf{a} + v\mathbf{b} + w\mathbf{c} \quad \text{and} \quad \mathbf{A}' = u'\mathbf{a} + v'\mathbf{b} + w'\mathbf{c},$$

and the angle φ between them is given by

$$\cos \varphi = \frac{\mathbf{A} \cdot \mathbf{A}'}{|\mathbf{A}||\mathbf{A}'|}.$$

Find the values of $\cos \varphi$ for (1) orthorhombic and (2) cubic systems.

Solution:

$$\mathbf{A} \cdot \mathbf{A'} = (u\mathbf{a} + v\mathbf{b} + w\mathbf{c}) \cdot (u'\mathbf{a} + v'\mathbf{b} + w'\mathbf{c})$$
$$= uu'a^2 + vv'b^2 + ww'c^2$$

$(\mathbf{a} \cdot \mathbf{b} = \mathbf{a} \cdot \mathbf{c} = \mathbf{b} \cdot \mathbf{c} = 0$ as angle between axial vectors for orthorhombic and cubic systems are 90°, i.e., $\alpha = \beta = \gamma = 90°$) and

$$|\mathbf{A}||\mathbf{A'}| = \sqrt{\{(u^2a^2 + v^2b^2 + w^2c^2)(u'^2a^2 + v'^2b^2 + w'^2c^2)\}}.$$

Therefore,

$$\cos\varphi = \frac{uu'a^2 + vv'b^2 + ww'c^2}{\sqrt{\{(u^2a^2 + v^2b^2 + w^2c^2)(u'^2a^2 + v'^2b^2 + w'^2c^2)\}}}. \qquad (B.9)$$

1. For orthorhombic, $\mathbf{a} \neq \mathbf{b} \neq \mathbf{c}$, so, $\cos\varphi$ is given by (B.9).
2. For cubic, $\mathbf{a} = \mathbf{b} = \mathbf{c}$ and so, (B.9) is changed into

$$\cos\varphi = \frac{uu' + vv' + ww'}{\sqrt{\{(u^2 + v^2 + w^2)(u'^2 + v'^2 + w'^2)\}}}. \qquad (B.10)$$

3. Two crystal direction vectors are given as $\mathbf{A}_{uvw} = u\mathbf{a} + v\mathbf{b} + w\mathbf{c}$ and in its reciprocal space $\mathbf{H}_{hkl} = h\mathbf{a}^* + k\mathbf{b}^* + l\mathbf{c}^*$, $\mathbf{H}_{hkl}$ is perpendicular to the *hkl* plane. If φ is the angle between them, then find the value of $\cos\varphi$ in terms of *uvw* and *hkl* for orthorhombic system.
   Solution:
   Taking the dot product of the two vectors $\mathbf{A}_{uvw} = u\mathbf{a} + v\mathbf{b} + w\mathbf{c}$ and $\mathbf{H}_{hkl} = h\mathbf{a}^* + k\mathbf{b}^* + l\mathbf{c}^*$, we get

$$\cos\varphi = \frac{(u\mathbf{a} + v\mathbf{b} + w\mathbf{c}) \cdot (h\mathbf{a}^* + k\mathbf{b}^* + l\mathbf{c}^*)}{|(u\mathbf{a} + v\mathbf{b} + w\mathbf{c})||(h\mathbf{a}^* + k\mathbf{b}^* + l\mathbf{c}^*)|}$$
$$= \frac{hu + kv + lw}{\sqrt{[(u^2a^2 + v^2b^2 + w^2c^2)(h^2a^{*2} + k^2b^{*2} + l^2c^{*2})]}}.$$

4. The zone axis *uvw* is defined by $\mathbf{A}_{uvw} = u\mathbf{a} + v\mathbf{b} + w\mathbf{c}$; all planes which contain the direction $\mathbf{A}_{uvw}$ are said to belong to the same zone. Deduce an expression of *hkl* planes which belong to this zone.
   Solution:
   $\mathbf{H}_{hkl}$ is the vector perpendicular to the *hkl* plane and as the planes contain $\mathbf{A}_{hkl}$, $\mathbf{H}_{hkl}$ will also be perpendicular to $\mathbf{A}_{hkl}$. Therefore,

$$\mathbf{A}_{hkl} \cdot \mathbf{H}_{hkl} = 0, \quad \text{i.e.,} \quad \varphi = 90°.$$
$$\mathbf{A}_{hkl} \cdot \mathbf{H}_{hkl} = (u\mathbf{a} + v\mathbf{b} + w\mathbf{c}) \cdot (h\mathbf{a}^* + k\mathbf{b}^* + l\mathbf{c}^*)$$
$$= uh + vk + wl = 0$$

is the required expression.

5. If γ^* is the angle between $\mathbf{a}^*$ and $\mathbf{b}^*$ in reciprocal space and is given by

$$\cos\gamma^* = \frac{\mathbf{a}^* \cdot \mathbf{b}^*}{|\mathbf{a}^*| \cdot |\mathbf{b}^*|},$$

then express $\cos\gamma^*$ in terms of α, β, and γ.

Solution:

Now, as $\mathbf{a}^*$ and $\mathbf{b}^*$ are expressed as

$$\mathbf{a}^* = \frac{\mathbf{b} \times \mathbf{c}}{V} \quad \text{and} \quad \mathbf{b}^* = \frac{\mathbf{c} \times \mathbf{a}}{V},$$

$$\cos\gamma^* = \frac{1/V^2[(\mathbf{b} \times \mathbf{c}) \cdot (\mathbf{c} \times \mathbf{a})]}{1/V^2|(\mathbf{b} \times \mathbf{c})||(\mathbf{c} \times \mathbf{a})|}$$

$$= [(\mathbf{b} \times \mathbf{c}) \cdot (\mathbf{c} \times \mathbf{a})]/[abc^2 \sin\alpha \sin\beta].$$

Now, from vector multiplication, $(\mathbf{b}\times\mathbf{c})\cdot(\mathbf{c}\times\mathbf{a}) = abc^2(\cos\alpha\cdot\cos\beta - \cos\gamma)$. Therefore, $\cos\gamma^* = (\cos\alpha \cdot \cos\beta - \cos\gamma)/\sin\alpha \sin\beta$, i.e., in terms of α, β, and γ.

6. Cesium chloride is simple cubic and has one CsCl per unit cell with Cs at 0 0 0 and Cl at $\frac{1}{2}\frac{1}{2}\frac{1}{2}$. Derive simplified expression for structure factor. Is there any systematic absences?

Solution:

Structure factor is

$$F_{hkl} = \sum_n f_n \exp[2\pi i(hx_n + ky_n + lZ_n)]$$

and for Cs at 0 0 0 and Cl at $\frac{1}{2}\frac{1}{2}\frac{1}{2}$ positions in the unit cell, we get

$$F_{hkl} = \sum f_{Cs} + \sum f_{Cl} \exp\pi i(h + k + l)$$

and the exponential is of the form $\exp(\pi im) = (-1)^m$. Now, for m is odd, i.e., hkl are all odd or any one is odd, so that $h + k + l$ are odd, then

$$F_{hkl} = \sum(f_{Cs} - f_{Cl}) \neq 0 \quad \text{as} \quad f_{Cs} \neq f_{Cl}.$$

Again, when $h + k + l$ are even,

$$F_{hkl} = \sum(f_{Cs} + f_{Cl}).$$

Therefore, there will be no absences and only intensity fluctuates between these two values.

7. Diamond is face-centered cubic with eight atoms per unit cell. Carbon atoms at 0 0 0 and $\frac{1}{4}\frac{1}{4}\frac{1}{4}$ positions and the other positions (six) are given by face-centering translations. Deriving simplified expression for the structure factor, find the rule for systematic absences and also the intensity of 2 2 2 reflection.

Solution:

$$F_{hkl} = \sum_4 \exp[2\pi i(hx_n + ky_n + lz_n)]$$
$$= 4\{f_c + f_c \exp[\pi i/2(h + k + l)]\}$$
$$= 4f_c\{1 + \exp[\pi i/2(h + k + l)]\}.$$

As this is complex, multiplying it with its complex conjugate, we get

$$F_{hkl}^2 = 16f_c^2\{[1 + \exp[\pi i/2(h + k + l)]][1 + \exp[-\pi i/2(h + k + l)]]\}$$
$$F_{hkl}^2 = 32f_c^2[1 + \cos \pi/2(h + k + l)].$$

When $(n = 0, 1, 2, 3, \ldots)$ $h + k + l = 4n$, $\cos 2n\pi = 1$ and $F_{hkl}^2 = 64f_c^2$; when $h + k + l = 4n + 1$, $\cos \pi/2(4n + 1) = 0$ and $F_{hkl}^2 = 32f_c^2$; and when $h + k + l = 4n + 2$, $\cos \pi/2(4n + 2) = -1$ and $F_{hkl}^2 = 0$; i.e., when hkl are all odd, $F_{hkl}^2 = 32f_c^2$; when mixed (one odd, two even) $h + k + l$ are odd $(1, 3, 5, 7, \ldots)$, $F_{hkl}^2 = 32f_c^2$; and when mixed (one even, two odd) $h + k + l$ are even $(0, 2, 4, 6, \ldots)$, the intensity will fluctuate between $64f_c^2$ and 0.

For 2 2 2 reflection, $h + k + l = 6$ and in $F_{hkl}^2 = 32f_c^2[1 + \cos \pi/2(h + k + l)]$, $F_{hkl}^2 = 32f_c^2[1 + \cos 3\pi] = 0$.

8. Zinc has hexagonal close-packed structure with Zn atoms (two atoms per unit cell) at positions 0 0 0 and $\frac{1}{3}\frac{2}{3}\frac{1}{2}$. Find the structure factor.

Solution:

Recalling

$$F_{hkl} = \sum_2 f_n \exp[2\pi i(hx_n + ky_n + lz_n)],$$

the structure expression comes out to be

$$F_{hkl} = f_{\text{Zn}}\{1 + \exp 2\pi i[(h + 2k)/3 + l/2]\}.$$

As this can be either real or complex, and so multiplying with its complex conjugate, we get

$$F_{hkl}^2 = f_{\text{Zn}}^2\{1 + \exp 2\pi i[(h + 2k)/3 + l/2]\}\{1 + \exp -2\pi i[(h + 2k)/3 + l/2]\}$$
$$= f_{\text{Zn}}^2\{2 + \exp 2\pi i[(h + 2k)/3 + l/2] + \exp -2\pi i[(h + 2k)/3 + l/2]\}$$
$$= f_{\text{Zn}}^2\{2 + 2\cos 2\pi[(h + 2k)/3 + l/2]\}$$
$$= f_{\text{Zn}}^2\{4\cos^2 \pi[(h + 2k)/3 + l/2]\} = 4f_{\text{Zn}}^2 \cos^2 \pi[(h + 2k)/3 + l/2].$$

Now, for $h + 2k = 3n$ and $l =$ even, $F_{hkl}^2 = 4f_{\text{Zn}}^2$; for $h + 2k = 3n$ and $l =$ odd, $F_{hkl}^2 = 0$; for $h + 2k = 3n \pm 1$ and $l =$ odd, $F_{hkl}^2 = 3f_{\text{Zn}}^2$; and for $h + 2k = 3n \pm 1$ and $l =$ even, $F_{hkl}^2 = f_{\text{Zn}}^2$.

9. Copper is a f.c.c. metal containing four atoms per unit cell at positions 0 0 0, $\frac{1}{2}\frac{1}{2}0$, $\frac{1}{2}0\frac{1}{2}$, and $0\frac{1}{2}\frac{1}{2}$. Find the systematic absences.

Solution:

Recalling

$$F_{hkl} = \sum_4 f_n \exp[2\pi i(hx_n + ky_n + lz_n)],$$

$$F_{hkl} = \sum f_{Cu}\{1 + \exp 2\pi i(h + k)/2 + \exp 2\pi i(h + l)/2 + \exp 2\pi i(k + l)/2\}.$$

Now, for *hkl* being either all even or all odd all the exponential terms will be equal to 1, then $F_{hkl} = 4f_{Cu}$ and when *hkl* are mixed, $F_{hkl} = 0$.

10. Tungsten is a b.c.c. metal having its atoms at positions as 0 0 0 and $\frac{1}{2}\frac{1}{2}\frac{1}{2}$. There are thus two atoms per unit cell. Find the reflection conditions.

Solution:

Recalling

$$F_{hkl} = \sum_2 f_n \exp[2\pi i(hx_n + ky_n + lz_n)],$$

$$F_{hkl} = \sum f_W(1 + \exp \pi i[h + k + l]).$$

When $h+k+l$ are even, $F_{hkl} = 2f_W$ and when $h+k+l$ are odd, $F_{hkl} = 0$.

11. Sodium chloride (NaCl) is two f.c.c. lattices (as both Na and Cl are f.c.c.) intervened within resulting into the positions of Na and Cl in one NaCl unit cell as

$$\text{Na} = \begin{matrix} \frac{1}{2}\ \frac{1}{2}\ \frac{1}{2} \\ 0\ 0\ \frac{1}{2} \\ 0\ \frac{1}{2}\ 0 \\ \frac{1}{2}\ 0\ 0 \end{matrix} \quad \text{and} \quad \text{Cl} = \begin{matrix} 0\ 0\ 0 \\ \frac{1}{2}\ \frac{1}{2}\ 0 \\ \frac{1}{2}\ 0\ \frac{1}{2} \\ 0\ \frac{1}{2}\ \frac{1}{2} \end{matrix}$$

Solution:

Recalling

$$F_{hkl} = \sum_n f_n \exp[2\pi i(hx_n + ky_n + lz_n)]$$

and considering one position each from the Cl and Na atoms, we get for *hkl* mixed, $F_{hkl} = 0$ and for *hkl* unmixed, $F_{hkl} = 4[f_{Cl} + f_{Na} \exp \pi i(h+k+l)]$; and also when *hkl* are all even, $F_{hkl} = 4[f_{Cl} + f_{Na}]$ and when *hkl* are all odd, $F_{hkl} = 4[f_{Cl} - f_{Na}]$.

12. A unit cell of tetragonal system has similar atoms at positions as $0\frac{1}{2}\frac{1}{4}$, $\frac{1}{2}0\frac{1}{4}$, $\frac{1}{2}0\frac{3}{4}$, and $0\frac{1}{2}\frac{3}{4}$ in the unit cell. Find the reflection conditions from the following reflections: 110, 222, and 111.

Solution:

Recalling

$$F_{hkl} = \sum_n f_n \exp[2\pi i(hx_n + ky_n + lz_n)]$$

and putting the coordinates of the atoms, we get

$$F_{hkl} = f\{\exp 2\pi i(k/2 + l/4) + \exp 2\pi i(h/2 + l/4)$$
$$+ \exp 2\pi i(h/2 + 3l/4) + \exp 2\pi i(k/2 + 3l/4)\}.$$

For mixed indices like 110,

$$F_{hkl} = f\{\exp 2\pi i(k/2) + \exp 2\pi i(h/2) + \exp 2\pi i(h/2) + \exp 2\pi i(k/2)\}$$
$$= 2f\{\exp \pi i(h) + \exp \pi i(k)\} = 4f\{\exp \pi i\} = 4f.$$

For unmixed like 222,

$$F_{hkl} = f\{\exp 3\pi i + \exp 3\pi i + \exp 5\pi i + \exp 5\pi i\}$$
$$= -4f \quad \text{and} \quad F_{hkl}^2 = 16f^2.$$

For unmixed like 111 all odd,

$$F_{hkl} = f\{\exp \pi i(1 + 1/2) + \exp \pi i(1 + 1/2) + \exp \pi i(1 + 3/2)$$
$$+ \exp \pi i(1 + 3/2)\}$$
$$= f\{\exp 3\pi i/2 + \exp 3\pi i/2 + \exp 5\pi i/2 + \exp 5\pi i/2\}$$
$$= 0.$$

Therefore, $F_{hkl}^2 = 0$.

13. Derive an expression for the resolution of a Debye–Scherrer camera for two wavelengths having very near values, in terms of separation of positions (S). S is defined in Sect. 7.2.
Solution:

$$S/R = 2\theta \quad \text{and} \quad \Delta S/R = 2\Delta\theta, \quad \Delta\theta = \Delta S/2R,$$
$$2d \, \sin\theta = \lambda \quad \text{and} \quad 2d \, \cos\theta\Delta\theta = \Delta\lambda.$$

Now dividing both, we get

$$\frac{\sin\theta}{\cos\theta\Delta\theta} = \frac{\lambda}{\Delta\lambda},$$
$$\tan\theta\frac{\Delta\lambda}{\lambda} = \Delta\theta = \Delta S/2R.$$

Therefore, $\lambda/\Delta\lambda = 2R \tan\theta/\Delta S$.

14. What is the smallest value of θ for which the CrK_α doublet can be resolved? In a 5.73-cm radius Debye–Scherrer camera, the smallest separation between two lines should be of width 0.06.
Solution:
In the expression as above, putting the following values

$$\Delta S = 0.06, R = 5.73 \text{ cm}, \lambda \text{ for } CrK_\alpha = 2.2910 \text{ Å},$$
$$CrK_{\alpha 1} = 2.2897 \text{ Å}, CrK_{\alpha 2} = 2.2936 \text{ Å}, \text{ and } \Delta\lambda = 0.0039 \text{ Å},$$

we get

$$\theta = \tan^{-1}\left(\frac{\Delta S}{2R} \times \frac{\lambda}{\Delta\lambda}\right)$$

$$= 72°.$$

15. Index the following lines obtained from a cubic system in a powder pattern with CuK_α and their measured $\sin^2\theta$ values are given as 0.1118, 0.1487, 0.294, 0.403, 0.439, 0.691, 0.727, 0.872, and 0.981. Confirm its Bravais lattice.

Solution:

$2d\sin\theta = \lambda$ and for a cubic system,

$$d = \frac{a}{\sqrt{h^2 + k^2 + l^2}},$$

$$h^2 + k^2 + l^2 = \frac{4a^2}{\lambda^2}\sin^2\theta = c\,\sin^2\theta,$$

where c is the constant.

For the first reflection, there are three possibilities of hkl (1) 100, (2)110, and (3) 111. The constant c is calculated from each of these possibilities:

1. From 100, $c = 1/\sin^2\theta = 1/0.1118 = 8.9445$.
2. From 110, $c = 2/\sin^2\theta = 2/0.1118 = 17.889$.
3. From 111, $c = 3/\sin^2\theta = 3/0.1118 = 26.833$.

From second reflection, it must be from 200 plane for any of the cubic system and so, $c = 4/\sin^2\theta = 4/0.1487 = 26.899$ and this value matches with the third possibility for first line; it can be concluded that the cubic system belongs to f.c.c. lattice. The reflections can be serially indexed as:

$\sin^2\theta$ values	–	Belongs to
0.294	=	220
0.403	=	311
0.439	=	222
0.583	=	400
0.691	=	331
0.727	=	420
0.872	=	422
0.981	=	333

It can be verified that the values of constant c as above can only match with possibility (3) for the above reflections and not for any other reflections which are thus not possible.

Further Reading

1. W. L. Bragg – The Crystalline State, Vol. 1, A General Survey, George Bell, London, 1933
2. A. H. Compton and S. K. Allison – X-rays in Theory and Experiment, D. Van Nostrand, New York, 1935
3. M. J. Buerger – X-ray Crystallography, Wiley, New York, 1942
4. W. T. Sproull – X-rays in Practice, Mc Graw-Hill, New York, 1946
5. R. W. James – The Crystalline State, Vol.2, The Optical Principle of the Diffraction of X-ray, George Bell, London, 1948
6. N. F. M. Henry, H. Lipson, and W.A. Wooster – The Interpretation of X-ray Diffraction Photographs, Macmillan, London, 1951
7. H. Lipson and W. Cochran – The Crystalline State, Vol. 3, The Determination of Crystal Structure, George Bell, London, 1953
8. L. V. Azaroff and M. J. Buerger – The Powder Method in X-ray Crystallography, McGraw-Hill, New York, 1958
9. R. P. Feynman, R. B. Leighton, and M. Sands – Lectures on Physics, Vol. 1, (Chap. 52), Vol. 2, (Ch.30), Addison-Wesley, Massachusetts, 1963
10. W. L. Bragg and G. F. Claringbull – The Crystalline State, Vol. 4, Crystal Structure of Minerals, George Bell, London, 1965
11. J. B. Cohen – Diffraction Methods in Materials Science, Macmillan, New York, 1966
12. L. V. Azaroff – Elements of X-ray Crystallography, McGraw-Hill, New York, 1968
13. B. E. Warren – X-ray Diffraction, Addison-Wesley, Reading, Massachusetts, 1969
14. M. M. Woolfson – An Introduction to X-ray Crystallography, Cambridge University Press, Cambridge, 1970
15. M. J. Buerger – Contemporary Crystallography, McGraw-Hill, New York, 1970
16. A. J. C. Wilson – Elements of X-ray Crystallography, Addison-Wesley, Reading, Massachusetts, 1970
17. F. C. Phillips – An Introduction to Crystallography, Wiley, New York, 1972
18. H. P. Klug and L. E. Alexander – X-ray Diffraction Procedures, Wiley, New York, 1974

19. S. Kumar (Ed) – Liquid Crystals, Cambridge University Press, Cambridge, 2001
20. S. V. Smaalen – Incommensurate Crystallography, IUCr, Oxford Scientific Publishers, 2007
21. I am-Choon Khoo – Liquid Crystals, 2nd Edition, Wiley, New York, 2007

Index

Springer Series in
MATERIALS SCIENCE

Editors: R. Hull R. M. Osgood, Jr. J. Parisi H. Warlimont

Springer Series in
MATERIALS SCIENCE

Editors: R. Hull   R. M. Osgood, Jr.   J. Parisi   H. Warlimont